The Television Code

The Television Code

Regulating the Screen
to Safeguard the Industry

DEBORAH L. JARAMILLO

University of Texas Press   Austin

The author and the publisher gratefully acknowledge permission to use the following material: "Keep Big Government out of Your Television Set: The Rhetoric of Self-Regulation before the Television Code," by Deborah L. Jaramillo. © 2016. From *Production Studies, The Sequel!: Cultural Studies of Global Media Industries*, edited by Miranda Banks, Bridget Conor, and Vicki Mayer. Reproduced by permission of Taylor and Francis Group, LLC, a division of Informa plc.

Requests for permission to reproduce material from this work should be sent to:
    Permissions
    University of Texas Press
    P.O. Box 7819
    Austin, TX 78713-7819
    utpress.utexas.edu/rp-form

♾ The paper used in this book meets the minimum requirements of ANSI/NISO
Z39.48-1992 (R1997) (Permanence of Paper).

Library of Congress Cataloging-in-Publication Data
Names: Jaramillo, Deborah Lynn, 1976– author.
Title: The Television Code : regulating the screen to safeguard the industry /
    Deborah L. Jaramillo.
Description: First edition. | Austin : University of Texas Press, 2018. | Includes
    bibliographical references and index.
Identifiers: LCCN 2017048952 | ISBN 978-1-4773-1644-3 (cloth : alk. paper) |
    ISBN 978-1-4773-1701-3 (pbk : alk. paper) | ISBN 978-1-4773-1702-0
    (library e-book) | ISBN 978-1-4773-1703-7 (non-library e-book)
Subjects: LCSH: Television broadcasting—United States. | Television broadcasting
    policy—United States. | Television—Law and legislation—United States.
Classification: LCC HE8700.8 .J37 2018 | DDC 384.55/4430973—dc23
LC record available at https://lccn.loc.gov/2017048952

doi:10.7560/316443

For Rodolfo, my favorite historian

Contents

Illustrations

Abbreviations

ABC	American Broadcasting Company
ACLU	American Civil Liberties Union
ASCAP	American Society of Composers, Authors, and Publishers
ATS	American Television Society
BBC	British Broadcasting Corporation
CAB	Citizens Advisory Board
CAD	Continuity Acceptance Department
CBS	Columbia Broadcasting System
DOC	Department of Commerce
FM	frequency modulation
FMA	FM Association
FCC	Federal Communications Commission
FMBI	Frequency Modulation Broadcasters Incorporated
FRC	Federal Radio Commission
FTC	Federal Trade Commission
ICC	Interstate Commerce Commission
IRS	Internal Revenue Service
ITPA	Independent Television Producers Association
JCET	Joint Committee on Educational Television
LAB	Library of American Broadcasting

LOC	Library of Congress
MBS	Mutual Broadcasting System
MPPDA	Motion Picture Producers and Distributors of America
NAB	National Association of Broadcasters
NACP	National Archives at College Park
NAEB	National Association of Educational Broadcasters
NARTB	National Association of Radio and Television Broadcasters
NBC	National Broadcasting Company
O&O	owned-and-operated station
PCA	Production Code Administration
RCA	Radio Corporation of America
SCABRAT	Southern California Association for Better Radio and Television
SCIFCS	Senate Committee on Interstate and Foreign Commerce Subcommittee to Study and Investigate Radio, Telegraph, and Telephone Communications
SCRC	Special Collections Research Center, University of Chicago Library
STE	Society of Television Engineers
TAC	Television Advisory Committee
TBA	Television Broadcasters Association
TCRB	Television Code Review Board
UHF	ultra high frequency
VHF	very high frequency
WHS	Wisconsin Historical Society

Acknowledgments

When I wrote my first book, I committed an embarrassing error and neglected to thank Dr. Barbara Selznick at the University of Arizona. Barbara is the model of a serious scholar and compassionate mentor. As her student and research assistant, I learned in my master's program how to do the practical and intellectual work necessary to succeed in my doctoral program. Everyone at the U of A was amazing, but Barbara deserves special recognition here.

From its inception, this project has benefited from a constellation of key people who wanted to see it in print. I'll start with Jim Burr at the University of Texas Press. Jim has been nothing but positive and supportive throughout this entire process, and I thank my lucky stars that Bernie Cook introduced us to each other at the Society for Cinema and Media Studies (SCMS) conference. I owe many thanks to the two scholars who provided thoughtful and endlessly useful feedback on the first drafts of this project. Reviewing sample chapters and the full manuscript is an enormous investment of time and energy, and it makes all the difference when the reviewers believe in the project and want to make it better. They made me work, and the book and I are better for it.

I had a great deal of help particularly through the research phase of this project. A generous bit of funding from Dean Thomas Fiedler at Boston University's College of Communication, expedited by Assistant Dean Maureen Mahoney, supported my travel to multiple archives. And my department chair, Paul Schneider, granted me a research assistant when the historical research began in earnest. Once at these archives I was drawn into a mountain of work and several years' worth of thrilling investigation. I want to extend my gratitude to the archivists and support staff at the National Archives at College Park, the Library of American Broadcasting at the University of Maryland, the Wisconsin Historical Society, the University of Chicago Library's Special

Collections Research Center, and the Library of Congress Motion Picture, Broadcasting, and Recorded Sound Division.

The work of researching and the work of sorting through all that research would have taken much longer than it did had it not been for the dedication of the incomparable Catherine Martin. I recruited Catherine to join our master's program in film and television studies in 2010. And then I recruited her to assist with this book. I couldn't have asked for a better partner. I hope that when she lands her first tenure-track job, she too is able to find graduate students as smart and generous as she is.

The Film and Television Studies Program at BU has had quite a run of stellar master's students, and I have benefited from their enthusiasm for research. Chelsea Daggett contributed to document organization. Laura Brown, Jason Henson, Megan Hermida Lu, and Austin Morris leapt into action when I was drowning in work during the home stretch. They gave up part of their summer to help me, and I will not forget their selflessness. Many times I have found myself motivated by my grad students, so it is no surprise that when Austin Morris wanted to wander into the literature of broadcast regulation, my own secondary research kicked into gear. I also owe thanks to the hundreds of undergraduates who saw the Television Code and NBC standards gradually creep into my syllabi. Their responses to questions surrounding the FCC's powers and the industry's self-regulatory impulses have kept the subject matter fresh for me over the years.

The community of media scholars and professionals in Boston is tight-knit, and I have benefited tremendously from being able to drink with, complain to, and share ideas with these friends. At BU, Roy Grundmann, Mina Tsay-Vogel, Cheryl Lambert, Garland Waller, Debbie Danielpour, and Mary Jane Doherty offered love and solidarity and generally kept me afloat during the tenure-track process. Outside of BU, I found necessary solace in the Boston Media Collective. As more and more media scholars find institutional homes in town, the BMC has grown. As a result, social life here has provided much-needed respites from the type of nonsense we as academics are all familiar with.

This book grew out of an early bit of reconnoitering at the National Archives, where I found an odd folder in the FCC's papers about astrology. Then came several conference presentations. I am indebted to the scholars I have presented with and those I have seen present at conferences. Miranda Banks brought Jen Porst and I together on an SCMS panel back in 2013, and a solid connection was formed. Talking through my research with Miranda and Jen made the process easier and, yes, joyful.

Finally, I am blessed with a family who loves me despite my temper. Dora and Frank Jaramillo worked their asses off so I wouldn't have to. They endured racism, bigotry, and hard manual labor as children and teenagers. They also put themselves through college while raising two kids. They are an example I can never live up to, but I'll keep trying. Alba Nora Martinez, the mother of my husband (she hates when I call her *suegra*), has been a fountain of generosity. I wrote my dissertation on her old desk. I wrote the first draft of this book on another of her desks in her outrageous midcentury home in Tucson. Everything she touches emanates good writing vibes, and I am thankful for that. I have no words to describe the gift that is my husband Rodolfo. He knows how I feel.

The Television Code and the Trade Association

The Television Code is a fascinating yet dull document, full of the anxieties and consensus politics of the 1950s, the appeasements of a defensive commercial industry, and the standardization of quotidian business deals. The original Code, adopted in October 1951 and implemented in March 1952, underwent twenty-one revisions until the entire enterprise collapsed in 1983. Perhaps because the Code regulated both program content and stations' dealings with advertisers, it is not understood to be as controversial or dramatic as the Motion Picture Production Code. And perhaps because multiple radio codes preceded the 1952 Code, scholars consider it to be more of the same—an "inevitable" step toward industrial maturity.[1] For these reasons—and probably for a few more—only two book-length examinations of the Television Code exist, and sadly, they were never transformed into books. One is a doctoral dissertation, and the other is a master's thesis. Robert Shepherd Morgan's 1964 dissertation, *The Television Code of the National Association of Broadcasters: The First Ten Years*, studies the Code's components and administration and devotes less than ten pages to the political wrangling that led to its drafting.[2] Emile C. Netzhammer III's 1984 thesis, "Self-Regulation in Broadcasting: A Legal Analysis of the National Association of Broadcasters Television Code," asks if the Federal Communications Commission (FCC) was complicit with the Code's regulatory mission.[3] Both are valuable pieces of scholarship, and both inform this book. Neither project considers how inherently problematic the process of content standardization is.

Television's transition from invention to business followed a simultaneously typical and not-so-typical trajectory. In typical fashion, its proponents established a trade association—a protectionist strategy common to US businesses. Atypically, this trade association found itself grappling with diverse displays of creativity. When these associations—first the Television Broadcasters Association (TBA) and then the National Association of Radio and Television

Broadcasters (NARTB)—set about managing the ins and outs of local television nationally, they also found themselves initiating and being drawn into conversations about the content of television programs. The motion picture industry and its trade association, the Motion Picture Producers and Distributors Association (MPPDA), had endured these same conversations and succumbed to pressures from outside the industry.[4] But as a private enterprise that was never regulated by a federal agency, the movie industry struggled through battles with state censorship boards and staved off cries for federal censorship.[5] Television was always regulated by the FCC, and though the FCC legally could not guide or censor content, concerns about indecent or inappropriate programs hovered over license renewal hearings. As Jennifer Holt writes, "The history of broadcast content regulation is defined more by its struggles than by any coherent set of rules or guidelines."[6] Struggles between the government and industry sometimes found temporary or piecemeal resolutions in self-regulation. Under these circumstances the NARTB felt it was making a valuable contribution when it created the Television Code, a set of standards that trumpeted the industry's moral authority and naturalized commercial TV at a crucial point in the medium's development.

Studying the Television Code

To dismiss the Television Code as mere public relations is understandable. Industry codes generally serve this function, at least in part. However, this sort of dismissal belies the peculiar yet integral role of the trade association in crafting that code.[7] Mostly ignored by media scholarship, trade associations are similar to the Television Code: fascinating yet dull. Situated between the state and the industry but steeped in both sets of practices, self-regulatory trade associations offer a unique lens through which to examine how social groups, unified by common jobs, negotiate relationships with public and private power brokers. The multiple roles of television's trade association—protector of local stations, mediator, researcher, lawyer, lobbyist, and agitator—deserve attention, as they amounted to a culture of self-regulation that championed wholesome content for economic reasons.

By centering the trade association, this book also accesses the significance of the local, a realm too often marginalized by a fixation on the national networks. Local broadcasters and national networks both belonged to the trade associations discussed in this book, but the former clearly outnumbered the latter. By 1952, 108 television stations were operating in the United States. The reason for the low number was the FCC's license freeze. The freeze, which commenced in 1948, halted the approval of license applications while the FCC rectified a channel assignment system that had resulted in poor television reception. While sorting out the allocation of television channels, the FCC deliberated other related concerns, such as the adoption of a uniform color television system. Consequently, what should have concluded in a matter of months lasted for four years, and during this time the four networks in existence—the National Broadcasting Company (NBC), Columbia Broadcasting System (CBS), American Broadcasting Company (ABC), and DuMont—busily forged affiliations with the few stations in operation.

Thomas Krattenmaker and Richard Metzger cite "physical" and "economic" reasons for the existence of television networks.[8] The shape of the earth requires the placement of multiple stations in order for broadcasts to reach the nation's population. The more stations there are, the more eyeballs advertisers can reach, which means that more money flows into the advertiser-supported system of production and distribution. Furthermore, as William Boddy explains, in the 1950s network affiliation represented the most desirable scenario for stations' solvency in "all but the largest and smallest markets."[9] A station affiliated with one or more networks had access to popular, live, well-funded programs. For this reason, the interests of local stations tended to converge with those of the national networks. The wedge between the two was the fact that stations held licenses and networks did not, leaving the large companies technically unregulated by the FCC and the affiliated stations at the mercy of network programs that may or may not have been compatible with local sensibilities.

Although all network affiliates were stations, not all stations were network affiliates. Independent stations relied on alternative sources of programming, but like the affiliates, they depended on advertising dollars, owing to their commercial status. Local broadcasters paid dues and occupied leadership positions in the trade associations while negotiating their roles as independent stations or as network affiliates. Hierarchies emerged within the associations, with larger stations dominating the smaller, cash-strapped ones. All were led to believe that their voices carried equal weight. To look into the archives of the National Association of Broadcasters (NAB) is to discover the politics of station owners and managers and to ascertain how some stations held more sway with their network partners than others. To study these broadcasters is also to understand the persistence of the local—local mores, local tastes, local commerce—in a medium widely recognized as a national force. Marsha Cassidy's *What Women Watched: Daytime Television in the 1950s* positions local and regional broadcasters as competitors of the national networks in the realm of early daytime television. This book seeks to develop that story at the level of self-regulation.

Although I argue that the heart of the Television Code is the trade association, a reader could easily argue that each component of this book, each contributor to the Code's formation, is at the center of the story. *The Television Code* is organized to allow and even encourage that interpretation. As Matthew Murray writes, the Code's "adoption [. . .] was not an isolated endeavor."[10] The Television Code was written by the NARTB, but it was mobilized in part by the FCC, elected officials, and viewers. Using archival documents from the National Archives, the Library of American Broadcasting, the Library of Congress, the Wisconsin Historical Society, and the University of Chicago, *The Television Code* explores the ideological, institutional, and personal pressures directed at those driving the commercial television industry in the 1940s and early 1950s. I argue that the Television Code was the result of a complex set of negotiations not just between government regulators and the national networks—the two institutions that loom large in the history of television—but also between local stations, their viewers, elected officials, and, of course, the trade association that would go to great lengths to shield the commercial system from discursive threats, state action, and technological alternatives. If successful, the Television Code would symbolize the superiority of commercial TV and the authority of TV's steward: the NARTB.

Although it is easy to explain the essence of the television industry and the reasons for its reactions to perceived threats by summoning the word "commercial," we need to employ methods and pursue lines of inquiry that reject simple narratives. Michaela Meyer supports a methodology that "embrace[s] the nuances between text, audience, and production," a list I argue should include industrial and governmental regulators.[11] To that end, I employ a critical political economy framework that, in Eileen Meehan and Janet Wasko's words, does not celebrate but rather "contextualizes those individuals, working cohorts, companies, and markets within the ongoing development of capitalism."[12] The political economy approach provides the tools to deconstruct a document like the Television Code and examine its components, causes, and consequences. An analytical picture of television sketched by Eileen Meehan positions the shifting allegiances and agendas of "individuals, corporations, trade associations, non-profit organizations, and governmental entities" in the foreground and the "institutional structures, rules, policy processes, and political agendas that together constitute the state" in the background.[13] Operating within an ideology of unquestioning "commitment to a privately owned, for-profit industry that uses public property for strictly private ends," different agents debated and settled on a distinct role for television in the mid-twentieth century.[14] Reformers who questioned that commitment were outliers and were ultimately expelled from the conversation. While scholars like Murray rightly describe the Code as a tactic whereby the NARTB was able to "deflect criticism away from the structural aspects of commercial broadcasting" and focus attention on program content, I argue that the document and the processes that led up to its drafting actually wanted very little attention trained on content. The Code was, paradoxically, a very public way of attempting to render invisible the means and ends of TV. The industry wanted the country to welcome television into the mainstream while exonerating itself and its goals. Television needed to become natural and dominant—a technological hegemon not up for discussion or debate. The Code upheld a specific plan for television that kept intact a state-supported, commercial system of broadcasting. The point was not only to stave off censorship or shift a critical gaze elsewhere; the stakes were much higher than that.

The fact of a predetermined role for television moves any analysis of the industry into ideological terrain. Studying the interaction between politics, economics, and ideology requires a structural approach to industry research. Meehan introduces this approach as a more comprehensive method than the instrumental approach, which is concerned primarily with the social relations within different organizations, and the institutional approach, which departs from "personal ties" and focuses instead on the ties between public and private organizations.[15] Identifying the structural approach as one that provides "deep context," Meehan argues that historical accounts vary according to the levels of context reached by the researcher.[16] She advocates for implementing "multiperspectival approaches" and ultimately formulating a new approach that draws from instrumentalism, institutionalism, and structuralism. Meehan's approach informs this book's analysis of the overlapping social and institutional relations at work in early television.

Relevant to this study, too, are the critical interventions made in the field of policy studies. While this book does not claim to originate from a political science framework, it does concern itself with the ramifications of communications policy and the interpretation of that policy within the body of the Television Code. To that end, the contributions made by critical policy scholars are invaluable. Critical policy studies teaches us that policy and policy analysis are not neutral exercises. If policy analyses are divided into the categories of "traditional" and "critical," then the "traditional" mode can be understood as an empiricist and social scientific approach to policy, a "linear process devoid of value judgments that focuses on measurable phenomena—free of power struggles."[17] The "critical" mode of analysis is a multidisciplinary approach that incorporates critical theory to expose policy as value-laden and hegemonic. For example, a feminist critical policy analysis, such as the type undertaken by Katherine Mansfield, Anjalé Welton, and Margaret Grogan to study education policy, acquaints policy analysis with intersectional feminism. Their work is interested in "how gender, race, and class shape the entire policy process," and it exposes how supposedly "gender-neutral" policies actually "negatively impact women."[18] Feminist policy scholars have also moved beyond uncovering sexism in public policy to proposing "alternative, equitable policies."[19] Scholars such as Frank Fischer and Sarah Diem and her colleagues argue that critical policy studies is the truest form of policy analysis as it was articulated by policy studies' founder, Harold Lasswell.[20] Lasswell's ambition was a type of policy analysis that incorporates "context, group values, and the contestable nature of problem definition, research findings, and arguments for solutions."[21] The approach was never supposed to be isolated from an interrogation of core assumptions and values.

Indeed, the policy scholar Timothy Luke warns of the regressive consequences of theoretical and methodological boundary-watchers.[22] Opening up the field to multidisciplinary influences has been essential to breaking down its conservative rigidity. Fischer associates critical policy studies with a turn away from positivism and a turn toward leftist politics, "normative political analysis," and "ideological critique."[23] A critical analysis pushes traditional policy analysis to ask how "larger systems of meaning" inform both policies and policymakers.[24] Fueled by philosophies of power, hegemony, and discourse from Antonio Gramsci, Michel Foucault, and Ernesto Laclau, critical policy analysis shares with critical political economy major theoretical insights as well as a commitment to social justice. One facet of that commitment can be found in the study of exclusion. David Howarth, Jason Glynos, and Steven Griggs describe their particular strain of critical policy analysis—poststructuralist discourse theory—as a means to decode the relationship between "power and exclusion" in the creation of "any policy practice or regime."[25] The procession to the enactment of policy is ideological and accordingly obscures or excludes alternatives.[26] In an analysis of the Television Code and its underlying structures, we should underscore the ways in which noncommercial and educational alternatives were obscured or aggressively excluded in the discourse of standard-setting.

This book also joins the project that scholars like Thomas Streeter and Jennifer Holt have embraced—that of placing history in conversation with legal frameworks that both contain and enable the flourishing of specific models of media ownership, production, and use.[27] The Television Code is important because it sits at the intersection of industry, government, and audience, of regulation and production, and of commerce and culture. It also occupies a key temporal territory since it rests firmly within a transitional moment for broadcasting and because its prehistory and implementation span the entirety of the television license freeze, which lasted from 1948 to 1952. *The*

THE TELEVISION CODE

THE NATIONAL ASSOCIATION OF RADIO AND TELEVISION BROADCASTERS
1771 N STREET, N. W., WASHINGTON 6, D. C.

FIGURE 0.1. The Television Code (1952)
Adopted by the NARTB in 1951 and implemented in 1952, the Television Code promoted responsible programming and business practices. Courtesy of National Association of Broadcasters.

Television Code probes these intersections and moments, positioning the Code as a turning point in the early history of mainstream commercial television—a period during which the mouthpiece of a young commercial television system heralded nationalism and anti-Communism in order to drown out the voices of opposition and the proponents of alternatives.

The Television Code (1952)

The 1952 Television Code is a nine-page document that sought, in part, to interpret broadcast television's public interest mandate without referencing the mission the government laid out for broadcasting in 1927. The Code was a set of standards crafted by the industry and not by the government, or at least not *directly* by the government. The NARTB pledged fidelity to mainstream values that were intended to reflect positively on everything about TV: the programs and the tangle of people and businesses that made TV happen. Certainly, the association also wanted the Code to seem not like recommendations but like enforceable directives. Owing much to the prevailing radio code and the Motion Picture Production Code, television's set of standards told the country it was safe to make this technology a part of daily life. These were the rules, and the industry would follow them. The NARTB had everything under control.

Accordingly, the Code was broken down into the following sections: "Preamble," "Advancement of Education and Culture," "Acceptability of Program Material," "Responsibility toward Children," "Decency and Decorum in Production," "Community Responsibility," "Treatment of News and Public Events," "Controversial Public Issues," "Political Telecasts," "Religious Programs," eight additional sections pertaining specifically to advertising, and a final section on "Regulations and Procedures." In its first paragraph, the Code employs the "guest-host" analogy to explain television's "responsibility" to its home viewers.[28] In the second paragraph, the Code reminds readers that advertising makes television possible, but that the broadcasters—the license holders—are ultimately responsible for what goes on the air. That "responsibility for excellence and good taste in programming" necessarily extends to all participants in the production process, including "networks, sponsors, producers [. . .] advertising agencies, and talent agencies." The document proceeds to explain how broadcasters are expected to meet the educational and cultural needs of their audiences. The terms that stand out—"excellence," "good taste"—are not defined explicitly. In listing the industrial players, the Code slides the established way of doing commercial radio over to TV and superimposes that model onto the intellectual and moral health of the public.

The next section, "Accep   Program Material," contains most of the prohibitions and recommendations for content (see appendix A). Twenty-eight entries specify the terms of wholesome TV, and they range from the expected (antisocial behavior should not be portrayed as positive) to the relatively unexpected (no astrology, no frightening aesthetics).[29] The subsequent section— "Responsibility Toward Children"—redirects the prohibitions listed in the previous section toward the well-being of the youngest TV viewers and offers additional guidance involving the portrayal of kidnapping, suspense, and mystery.[30] The section entitled "Decency and Decorum in Production" lists four production-specific directives related to costumes, bodily "movements," suggestive camera angles, and sexually suggestive locations (see appendix B).[31] An extra instruction to avoid ridiculing a "race or nationality" appears in this section, though its lack of production specificity renders it out of place and more applicable to the general prohibitions.[32] Following a brief paragraph about broadcasters' responsibilities to their communities, the Television Code sets standards related to tasteful and accurate coverage of news, public events, controversial issues, political programming, and religious programming.[33] The next three pages of the Code meticulously outline acceptable advertising content and standardize commercial time limitations according to the length of the program.

The final pages focus on process and clarify the mission of the Code, which instructed stations "cooperatively to maintain a level of television programming which gives full consideration to the educational, informational, cultural, economic, moral and entertainment needs of the American public to the end that more and more people will be better served."[34] Television stations could subscribe to the Code for a fee voluntarily and upon doing so would be allowed to display the "NARTB Television Seal of Good Practice."[35] If the five-member Television Code Review Board (TCRB) determined that a subscribing station's programming or advertising violated the tenets of the Code, then the NARTB's Television Board of Directors had the authority to strip that station of its seal.[36] The process allowed for appeals and hearings, so the punishment was not as final as it appeared. Loss of the seal did not necessarily endanger the station's license; the power of the seal's presence or absence was merely symbolic. Appealing to a station's sense of ethics via the seal and policing station behavior via program monitoring were the only legal ways to maintain the Code at all. These were the "equivalent of enforcement," as former NARTB president Justin Miller noted.[37] Methods of enforcement that carried tangible consequences might be construed as anticompetitive behavior. Regardless of its ethical or legal connotations, the Seal of Good Practice, which will be analyzed further in this book's conclusion, announced that the work of standardization had made the screen safe for public consumption.

Standards and Standardization

Regulation is, for Michael Clarke, less about creating rules and more about the pursuit of "order" in a sufficiently unruly field.[38] Government bodies may attempt to create order from above, or private bodies may attempt to mend ruptures themselves. Both scenarios can also transpire within one industry, which is what we see in the field of broadcasting. Government regulation will be discussed at length in chapters 1 and 5, so for the moment we will focus on self-regulation, of which the Television Code is one example. When an industry forms a trade association and self-regulates, it "exerts control over its own membership and their behavior."[39] Trade associations, composed of different people who ostensibly are competing with one another, are "forced to accept at a collective level the need to keep the economic show on the road."[40] One way to do this is through standardization, a process that takes on a set of complicated meanings when applied to the television industry.

As a culture industry in which advertising is the main source of immediate funding, broadcast television yields no traditional, tangible product that a producer exchanges with a consumer.[41] At the national level, a network sells audiences (in the form of ratings) to advertisers. Commercial broadcast audiences pay for television programs not directly but by buying the products advertised during the programs. These processes occur in what Streeter calls a "simulation," following the work of Jean Beaudrillard.[42] In this "indirect market mechanism," the advertiser is the audience surrogate, the ratings are the "administrative stand-ins for the market," and statistics and formulas replace the traditional money exchanged for goods.[43]

The simulation that Streeter theorizes cannot hold absolutely for the trade association. For that organization, the product must be, at least partially, the television program. When the trade association adopts standardization as a policy, programming must be understood in ways that apply to the marketplace rather than to creativity. Originality, for example, can be a problem. Trade associations need to mitigate the disturbance that might ensue from the circulation of too many products that stray from the norm. A 1925 publication explains the problem of unharnessed variety. When the producers of those products are "uncoordinated" and cannot determine consumer demand, "a vast flow of unstandardized products tends to emerge which do not appear to serve any significant economic purpose."[44] The trade association exists to facilitate a degree of cooperation between businesses "to achieve greater standardization and simplification of products."[45] The aims of standardization are to lower costs, "enhance the ease with which goods are duplicated and distributed," and "safeguard the consumer against uncertainty."[46] One danger of such standardization, according to Clarke, is the attendant pressure to form a cartel. If a product or its technology becomes standardized, companies "find it impossible to compete by innovation, and competition, if pursued, becomes mutually destructive."[47] Television broadcasting, subject to the standardization of its image quality in the early 1940s, was ushered in the 1950s toward the standardization of its content by both government and private regulators.

Standardization thrives in the capitalist logics of media production. Genres, as discussed by Thomas Schatz, Jane Feuer, Steve Neale, and Murray Forman have performed a standardizing function in the radio, film, and television industries.[48] What I find distinctive about standardization in the case of the Television Code (and the radio codes, which will be discussed in the next chapter) is its enactment by a body completely separate from creative labor. Divorced from the process of creation but committed to confining it within a system of rules, the NARTB activated a process that attempted to limit even the emergence of certain genres. Creative workers and trade associations negotiate the "tension between similarity and difference" for different reasons.[49] For creative workers, uniqueness can (and must) have a place within a system of conventions. For the trade association, standardization can reduce competition between businesses and uniqueness can trigger more competition. A competitive system can result in a true diversity of products that "narrows the market" (think niche audiences), but this process was believed to cause "violen[t] fluctuations in effective demand" in the early twentieth century.[50] The trade association's stabilizing influence, then, was its pursuit of standardization and less competition.

As the rise of original dramas and comedies on cable television in the late twentieth and early twenty-first centuries has demonstrated, deviation from broadcast formulas can spur tremendous competition for market share between broadcast networks and cable channels. The niche audiences created by cable television, which different tiers of cable service further winnowed, were evidence of content diversity if not diversity of corporate ownership. From the perspective of the twenty-first-century television landscape, it is important to recognize that the restrictions on content implemented by the Television Code in 1952 did not just establish standards of decency; they worked to standardize program themes as well as stylistic choices. While the Code and broadcast network standards departments relaxed these restrictions over time, a minimum engagement with standards was necessary for license holders, which were and continue to be the local television stations. As Harvey Levin argues, whether or not the Code was effective, it nevertheless led broadcasters to engage in an "interfirm product agreement" that "maintain[ed] certain product standards through inclusion and exclusion."[51] The two uses of "standards" here—one signifying the establishment of a baseline product formula and the other signifying the establishment of a baseline of moral quality—combine in the form of the program, the visible and audible demonstration of the industry's ethics and morals.

The conflation of moral standards and product standardization within the body of the television program requires further investigation into the role of ethical conduct in business enterprises. According to one early examination of trade associations, "group opinion," "the dread of social disapprobation," and "the prospect of receiving the approval and acclaim of others" factored into the codification of "moral rules" for businesses.[52] To quote this early study:

> A business code of ethics [. . .] is a general statement of the course of conduct in relation to the common transactions of a given trade or industry which is consistent with prevailing moral principles as they are interpreted by those engaged in the particular business. The adoption of a code of ethics involves an implicit recognition of the fact that sometimes the methods of "successful" business operation are at variance with the moral standards and long-run interests of the group as a whole and hence must be tempered by the infusion of ethical restraints.[53]

The Code's existence implied that racier or purely entertaining programs might yield higher ratings—a more "successful business operation"—but pressure from viewers, Congress, and the FCC helped to create an atmosphere in which racier programs would be "tempered" by standards. For Levin, the Code was a "defensive" strategy deployed in pursuit of someone's version of a "higher" cultural standard.[54] That standard, I argue, was designed to normalize commercial TV and to insist that commercial TV implicitly operated in the public interest, though the short-term goals certainly included "keep[ing] the federal camel's nose out of the industry's tent" and pleasing—or at least pacifying—viewers.[55]

Some members of the public had been quite vocal about what they perceived to be the moral failings of radio and, later, television. Thomas Doherty describes viewer feedback as even "more stringent than the regulatory authority of the FCC."[56] In his article on the regulation of sexuality in early radio, Matthew Murray examines both the content and consequences of Mae West's infamous Garden of Eden skit on *The Chase and Sanborn Hour* on NBC in 1937. Religious organizations, women's groups, and the press condemned the sketch for its "suggestive dialogue," but Murray takes a closer look at the outrage by analyzing the interaction between sound and mind. "Radio comedians and scriptwriters," he notes, "relied upon language's complexity and interpretive openness to aurally titillate listeners."[57] How to control for the "variables" of "performers' histrionic inflections and the audience's various socially influenced receptions" created a "procedural crisis" for the National Broadcasting Company's (NBC's) Continuity Acceptance Department (CAD). At issue here were form and formal variations that resisted the standardization of content. In radio, sonic variables impeded the strict imposition of standards. In television, the addition of sounds to images may have slowed viewers' runaway imaginations, but it contributed to a new type of communication (visuals plus sounds plus liveness) that required additional layers of oversight to quiet the loud expressions of moral panic.

Standard-setting on television sought to appease angry viewers and in the process compromised the possibility of competition via program diversity. As Stanley Besen and his colleagues explain, "Reliance on competition among stations, networks, and program suppliers for consumers' patronage reflects the belief that, in general, the mix of programs that results from this competition will correspond closely to that mix desired by viewers."[58] Restricting content—fixing the themes and ideas that writers can write about and television stations can broadcast—stifles the program options that Besen and his coauthors imagine.

So we return to the problem undergirding this study. Just as television is not "a toaster with pictures," as former FCC chair Mark Fowler proposed, television programs are not everyday consumer goods.[59] In the traditional functioning of the commercial television industry, programs are not the real products at all. Audiences are the real products. However, if the trade association is conditioned to view television programs as mass-produced goods, how does it ensure "quality"? How does a regulating body, staffed by noncreatives, standardize creativity? How did the engineers of the Television Code imagine programs interacting with their foundational concern: the preservation of the commercial broadcasting system? And finally, how can we think of standard-setting as a process of normalizing and making invisible an anticompetitive industry intent on fighting alternatives and silencing opponents? As the following chapters offer answers to these questions, they also tackle conflicts and cooperation across social and institutional lines. Everyone, powerful or not, had questions about what television should be, what it should do, and how it should do it. The Television Code tried to offer the definitive answers.

Chapter Breakdown

We cannot begin to approach the Television Code without first becoming acquainted with its antecedents in radio. The first part of chapter 1 explores how the content of radio messages factored into government regulation of early radio. The second part of this foundational chapter then reviews how the NAB folded content into its own regulation of radio. The numerous radio codes that preceded the NAB's involvement in television established a precedent for the Television Code and exemplified the association's approach to a creative and commercial medium.

Chapter 2 excavates the little-known story of the Television Broadcasters Association, television's first trade association. Unexamined by scholars, the TBA launched quietly in 1943 and struggled to attract enough membership to make it a viable companion to the NAB. Since the NAB had not considered television seriously, the TBA was the lone national voice for television interests in the mid- and late 1940s. This chapter examines the brief life span of the TBA and its takeover by the NAB in 1951, at which time the NAB temporarily changed its name to the National Association of Radio and Television Broadcasters. The strife between the two trade associations raises questions pertinent to commercial television's path.

Chapter 3 extends the focus on trade associations by examining the ways in which the TBA and the NAB envisioned restrictions on program content. NBC already had its Continuity Acceptance Department in place for the new challenges posed by television. But network policies do not necessarily represent the needs or desires of local stations, which were the primary members of the broadcast trade associations. TBA conference proceedings from 1944, 1946, and 1948 indicate clearly how TBA members were beginning to contemplate the ramifications of censoring for taste and decency. Although the TBA grappled with its own ideas about a code well before mainstream TV launched, once the NAB embraced television—dissolving the TBA in the process—the journey to the Television Code was a short one. Internally the NAB proceeded toward a code to protect its television membership, but externally the NAB transformed that pragmatic approach into a series of flamboyant tirades against the government.

Every participant in this narrative—the TBA, NAB, FCC, networks, and Congress—regularly invoked the well-being and complaints of viewers. Who were these vocal audience members, what were they complaining about, and to whom were they directing those complaints? Chapter 4 shifts the book's gaze from the industry's representatives to the industry's consumers and analyzes the major and minor concerns raised in their letters to the FCC. Housed in the National Archives, the surviving letters from the late 1940s and early 1950s range from simple pleas for decency to lengthy diatribes about television's pernicious effects on the nation. Regardless of their content, all of the complaint letters (there were a few letters of praise) shared the belief that the FCC could censor TV and that broadcasters were answerable to both the public and the government.

The government's involvement in the Television Code's development took two forms. The first form was the FCC, which had been receiving and responding to complaint letters since its creation in 1934. This seven-member commission, created by and answerable to Congress, found itself in a complicated position when television content became a pressing issue in the national conversation. Inundated with the letters discussed in chapter 4, the FCC behaved partly as a regulator and partly as a middleman between the NAB and Congress. Chapter 5 first examines the odd position of the FCC by exploring its simultaneously prickly and cozy relationship with the NAB. The FCC's power as an administrative agency—a regulatory body with no constitutional footing—paired with its considerable weakness as an overworked and underfunded agency, solidifies its ambiguous role in the Television Code narrative.

The government's involvement also took the form of Senator William Benton, a television reformer and whipping boy for the NAB. Multiple senators and representatives were vocal about television content, but Benton kept a high profile and was already known to the industry as a prominent advertising man from the early days of radio. Chapter 6 analyzes the battle between Benton and the NAB—a battle that allegedly hinged on Benton's plan to create a National Citizens Advisory Board for Radio and Television. Using trade publications, NAB papers, and Benton's papers from the Special Collections Research Center at the University of Chicago, this chapter argues that Benton posed a legitimate threat to the NAB, but not because he wanted a board of citizens to provide annual assessments of radio and television shows. Benton unsettled the NAB because he was an enthusiastic proponent of subscription television and educational television, both of which called into question the commercial system's superiority and attentiveness to the public interest. As a threat to the advertiser-supported model, Benton had to be stopped.

After the industry adopted the Television Code, it faced congressional hearings on radio and television content. From newspapers to trade magazines to the halls of Congress, the Television Code was on people's minds, and the NARTB had a great deal to do in the way of promotion and persuasion. The concluding chapter therefore revisits the book's major themes by looking at the work the association still had to undertake to convince Congress, the public, and stations that the Code meant what it said, even though it really meant so much more.

1943–1952

The media landscape was profoundly transformed during the brief time period covered by this book. Television became a national hobby, and soon enough it became a scapegoat for any number of social ills: juvenile delinquency, intellectual laziness, physical inactivity. Television acquired a reputation for being an "idiot box" and, worse, an artistic desert. In a 1952 *Life* article, RCA chairman David Sarnoff openly badmouthed TV: "We in television already have Hollywood licked in mediocrity. They can't touch us there."[60] A number of factors paralyzed this new medium in its infancy. One was, as Philip Sewell notes, the fact that "television programming was imagined and realized largely outside public view," and another, I argue, was the rush to create standards of practice in the service of a naturalized and static commercial practice rather than a dynamic and creative one.[61] The consequences for creative practice stemmed from conservative attitudes about decency that had been marshaled at different times by the NAB, NBC, and Hollywood's Production Code Administration (PCA). The Television Code was just one more list of "don'ts" added to the pile.

But its circumstances were different. Unlike the various radio codes in existence since the 1930s, the Television Code was fully formed with mechanisms in place to reward stations that subscribed to it. Unlike Hollywood's Production Code, the Television Code deferred to democracy and capitalism—not morality—as ultimate authorities. And unlike NBC's standards, the Television Code applied to all member stations, not just the affiliates of one network. Each of these restrictive documents emerged at specific times for specific reasons; each amplified a particular voice responding to a distinct set of pressures. This is the story of the pressures that led to the Television Code.

CHAPTER 1

Regulatory Precedents before Television

The Government and the NAB Experiment with Radio

Just as the television industry has deep roots in the radio industry, so too does early TV regulation draw from the public and private regulation of radio. The history of radio regulation reveals that the government has always involved itself in content. As much as the industry might want to escape it, legislation created a regulatory regime that could be swayed, fought, or appeased, but never destroyed. The government's approach to the changing circumstances and uses of radio established a template for dealing with television. Revisiting this approach helps to clarify the expectations the NAB had of the FCC once it began to plan for television in a serious manner.

After reviewing key moments in radio regulation, this chapter shifts to the industry's method of regulation, itself one indication that the NAB would treat broadcasting as any other industry. A lack of innovative thinking initiated responses to government regulation that favored networks, network affiliates, and advertisers rather than creative labor. In fact, creativity and artistry, while useful as buzzwords for the industry, were compromised consistently by decisions bearing on content. Predictably, creative laborers and artists were rarely (if ever) included in deliberations about rules and standards. The government and trade association managed the substance of radio and television messages as though they could be manufactured and controlled without interference or input from the talent. The regulatory mechanisms the FCC and NAB employed differed according to their respective powers and limitations, but they both sought to construct a stable space for a system aligned with corporate liberal ideas. The public and private regulations established for early radio kick off the story of the standardization and naturalization of television.

The Government and the Shifting Terrain of Early Radio

Prior to 1910, professional and amateur wireless interests had opposed government regulation of radio's prebroadcasting form. The lack of cohesion among wireless industry players, the relative unimportance of wireless regulation, the technology's inchoate state, and the absence of similar communication regulation left wireless relatively untouched by legislators until, as Susan Douglas states, the well-being of people became an issue.[1] Although in 1910 the Mann-Elkins Act granted the Interstate Commerce Commission (ICC) some power to regulate wireless telegraph carriers, wireless underwent intense government scrutiny only after two vessels collided at sea.[2] The result was the Wireless Ship Act of 1910, which mandated that wireless receivers and transmitters be installed on all ships entering or leaving the United States. Douglas argues that this piece of legislation positioned Congress as a "protector," a guardian of immigrants coming to the United States by sea.[3]

This was the first of many instances in which Congress preferred to let the experts resolve their own issues as much as possible before acting on them. At one level, Congress was correct to allow inventors and engineers to sort out interference and other problems associated with the electromagnetic spectrum. At another level, Congress backed away cravenly from the complications of communication, public airwaves, and private ownership. Douglas explains that a central question—"What criteria should Americans use to assign and protect property rights in the spectrum?"—vexed legislators then and continues to inspire arguments.[4] The question would be addressed but not altogether settled in the form of a government-issued license. After the *Titanic* disaster, the push for greater government regulation, a trend emblematic of the Progressive Era, resulted in the passage of the Radio Act of 1912. With the help of the US Navy, the government issued licenses to institutional and amateur wireless operators. The Radio Act of 1912 systematized core beliefs about how the airwaves should be perceived and who could best manage them within the confines of that perception. According to Douglas, the 1912 act converted frequencies into property, privileged institutions over individuals, and placed great faith in the ability of "consolidated institutions" to look after the public interest.[5]

Charles Tillinghast notes that while some historians consider the 1912 act to have been ineffectual, the law was, in fact, of supreme importance because the federal government inserted itself into radio via the commerce clause of the US Constitution.[6] As Tillinghast notes, "No consideration was given to the possibility that some other part of the Constitution, specifically, the First Amendment, might be relevant and that use of wireless might also involve 'speech.'"[7] To illustrate his point, Tillinghast distinguishes between the regulation of a wired telegraph service like Western Union and that of a wireless communication like radio. The regulation of Western Union did not touch the content of the messages, whereas licensing wireless operators "was a limitation on the right to employ this method of communication at all."[8] In other words, while the government would not participate in the radio game, it would pick the players. As Streeter asserts, the 1912 act would encase spectrum-related issues within a legal framework adjudicated by federal administrative agencies.[9] By issuing licenses, the government automatically insinuated itself into content. Douglas writes that the lucky recipients of licenses had convinced the government that they could commit the necessary resources and that they had something important to say.[10] Before the fact of broadcasting transformed the significance of the license, the government needed to manage its oversight of point-to-point radio.

Herbert Hoover, appointed secretary of commerce by President Warren G. Harding, became the de facto regulator of radio and began issuing broadcast licenses in 1921. The extent of his power was ill defined. When he decided not to renew one particular license because of concerns about signal interference, the US Court of Appeals in Washington, DC, found that Hoover had overstepped his authority.[11] According to Hugh Aitken, until 1927 the issuing of radio licenses "was merely a matter of registration."[12] That Hoover tried to block access was not a matter of course; rather, it was an attempt to control an "overcrowding" of the airwaves.[13]

Before point-to-point communication transitioned to broadcasting, radio companies like RCA felt that they had the technology in hand and were quite comfortable with the minimal role that the Department of Commerce assumed in fulfilling its end of the licensing arrangement. Aitken explains that the displacement of spark transmitters by vacuum tubes and the adoption of these tubes for the purposes of broadcasting signals to many listeners at once disturbed the relative harmony that followed the Radio Act of 1912.[14] Eventually more stations wanted to broadcast than was feasible under the existing technological and institutional limitations, which Aitken explicates in detail. To attempt to manage this situation, which the 1912 act had not anticipated and could not accommodate, Hoover held four radio conferences from 1922 to 1925. The conferences were what Louise Benjamin calls "cooperative efforts" led by the government but welcomed by the industry.[15] Streeter characterizes the conferences as acts of "Hoover associationalism"— a "mode of interaction" between business and government that was particular to Hoover's operating style.[16] According to Robert McChesney, as Hoover looked to the powerful participants to offer guidance, the conferences opened the door to self-regulation.[17] But McChesney disagrees that the network- and advertiser-dominated medium that radio became was already determined before 1927.[18] The "economic instability" that characterized radio in the 1920s, as well as Hoover's pro–public service stance made apparent at his radio conferences, gave no indication that nonprofit and noncommercial stations would have to fight to stay on the air after 1927.[19]

The conferences, attended by representatives of like-minded and even adversarial interests, tackled both new and recurring issues, including signal interference, licensing, advertising, and censorship. Streeter argues that the conferences were an ideological exercise, a process of delimiting conversation and naturalizing ways of doing and controlling radio.[20] One such exercise involved treating licensing as a technological concern isolated from content. Content was on conference participants' agendas, however, and it rapidly became apparent that broadcaster responsibility and program subject matter were tethered to one another. Benjamin writes that at the 1924 conference RCA, AT&T, and Westinghouse representatives advanced specific ideas about "quality" that revolved around "public interest" and "program acceptability."[21] These early conversations illustrate what Sewell describes as the "productive nebulousness" of "the discourse of quality."[22] For Sewell, this discourse connects content to "cultural values" while also enabling "a seemingly coherent system of cultural and economic exchange."[23] And while at any point in time, he concedes, institutions may deploy the discourse of quality differently, "the notion that quality is recognizable and desirable is crucial to its cultural utility."[24] The recognizability of quality is also crucial to its economic utility. The use of the term "quality" in economics is similarly vague but essential. Carlo Scarpa emphasizes that, far from being "a banal issue [. . .] quality is an attribute of a good of which all consumers prefer more rather than less."[25] In broadcasting, even this definition is opaque and unhelpful. Understandably, the nascent discursive construction of quality at the radio conferences did not develop into coherent or concrete action. The major companies stopped short of codifying anything that would necessarily invite government intervention. They agreed not to hand the government a reason to set its sights on content.

Each of Hoover's radio conferences attempted to push legislators to take up the problems of radio, and finally the confluence of growing radio "chaos" and Hoover's regulatory impotence assured by *US v. Zenith Radio Corporation* forced Congress to act.[26] With the passage of the Radio Act of 1927, which the NAB and other commercial radio interests supported, radio would finally be overseen by a dedicated independent government agency with the power to decide who had access to the spectrum (although the Radio Division of the Commerce Committee maintained some regulatory authority until 1932).[27] Significantly, Congress did not treat the newly created Federal Radio

Commission (FRC) as a permanent institution. The legislative branch of government needed to reauthorize the FRC yearly and only allowed it a minimal budget.[28] Not until 1929 did Congress grant the FRC continuing license authority without annual review.[29]

Endowed with the authority to review allocation decisions made by the Radio Division and to issue licenses, the FRC did not treat all broadcasters equally. The commission abided by and interpreted what Allison Perlman asserts was an ideological endeavor to "privilege the national over the local, the commercial over the non-commercial."[30] Furthermore, the FRC's mandate "to allocate licensees on the basis of which prospective broadcasters best served the 'public interest, convenience, or necessity'" placed content on its agenda.[31] But the FRC's power to assign frequencies was framed as an "engineering" exercise rather than an evaluative process involving quality or ideology.[32] Because engineers linked financial abundance to public service, "the best slots [were allocated] to the best capitalized stations."[33] Consequently, the "vision of broadcasting" established by government and private entities endangered noncommercial broadcasters, resulting in their near-complete marginalization by 1935.[34]

The relationship between the government and private industry, which will be explored in greater detail in chapter 4, is often described as cozy and friendly. Scholars correctly point to the FRC's General Order 40 as evidence that the commission privileged commercial stations over noncommercial stations in license hearings. Under General Order 40, which was a channel reallocation plan instituted in 1928, the FRC's engineering staff made room for forty high-power, clear channel stations. The goal of the plan was to improve reception, but in pursuing it the commission had created a system that "favor[ed] those broadcasters with the best technical equipment."[35] In another blow to small broadcasters, the FRC labeled noncommercial stations "propaganda stations" and denied them the same access to the spectrum that their commercial counterparts enjoyed.[36] A frequency-sharing scheme also limited the airtime of less "worthy" stations, which often were the stations deemed to be propagandistic.[37] After one year, the rule had reduced the number of stations on the air by one hundred.[38] The result was a large-scale disenfranchisement of noncommercial broadcasters. By 1931, NBC and CBS supplied "nearly 70% of US broadcasting," and this dominance paralleled the rise of direct advertising, a tactic shunned at Hoover's radio conferences.[39] This type of regulation has led scholars to position the government regulator as "the steward/patron of major corporations and big broadcasters."[40]

Benjamin argues that as one radio conference followed the other, "a close-knit interrelationship between regulator and regulated arose."[41] Holt's idea of "structural convergence" is visible in what Benjamin refers to as the "associative state" that resulted from the industry-government interactions between the 1912 and 1927 acts.[42] McChesney likewise chronicles the ways in which both elected officials and appointed commissioners helped commercial broadcasters maintain the status quo after 1927. With one exception, the FRC was populated with commissioners who had no wish to raze the commercial system. According to McChesney, the FRC "allow[ed] the industry to determine the nature of broadcast regulation as much as possible, regarding it as an ally."[43] Commercial broadcasters did their best to control access to the spectrum, swaying politicians with arguments about Americanism and free speech and feeding their expertise and well-funded research to the commission's decision-makers. An integral but overshadowed participant in the commercial broadcasting juggernaut was the National Association of Broadcasters, the self-regulatory voice of the industry.

The National Association of Broadcasters and the Turn to Self-Regulation

Although the creation of a trade association for broadcasters was not in Hoover's plans, the NAB did form in the midst of the radio conferences. According to David Mackey, approximately eight "very discontented broadcasters" met in Chicago in 1923 to discuss music royalty payments and in doing so "realized that a trade association was vital to the health of this new and expanding industry."[44] In April of that year, a committee of seven men recommended the formation of a trade association to be headquartered in New York City.[45] The romantic version of the formation of the NAB, told by its president, Harold E. Fellows, in a 1952 speech, lists five reasons those early radio men looked to each other for support.[46] The first was "to oppose certain unseemly demands being made upon them for music licensing fees." The second was to ensure that stations were licensed in an "orderly" fashion. The third was to "resist" harmful government action. The fourth was to establish a public relations mechanism in order to deflect "public criticism." And the fifth was to harness advertising for the benefit of the industry.

Fellows proceeded to characterize the NAB's history as a series of stabilizing efforts. The NAB's "Development Stage," between 1922 and 1942, was driven by "pioneers" who sought "stability, expansion, and protection." World War II dominated the second stage, when the NAB focused its attention on the war effort. The third stage, or the "reorganization period," transpired when the NAB struggled with a series of "vexing problems"—the impressive growth of radio stations, the blossoming of television, the entrance of

FM—and ultimately triumphed. The fourth stage, the current stage at the time of Fellows's speech, was characterized by a unified front. "Stations, networks, manufacturers, and special services have closed ranks," Fellows remarked. The unification of radio and television within one organization made such a stalwart defense possible. Fellows referred to that unification as a "noble experiment," but it was clearly an economic boon. The NAB (renamed National Association of Radio and Television Broadcasters—NARTB—in 1951) stood to benefit substantially from the expansion of television service once the license freeze ended. As of July 1, 1951, 107 television stations were in operation, and 415 TV station applications awaited approval from the FCC.[47] Even in the midst of the freeze, over 14 million television sets were in US households.[48] By 1952, the year the FCC ended the freeze, the NARTB would be poised to oversee television's rapid growth.

If we return to that pioneering "development stage," as Fellows described it, we can begin to understand why self-regulation was so important to the industry. Erik Barnouw describes the NAB as the "standard bearer in many battles" that broadcasters would wage against competing interests.[49] Yet, of the multiple major players in and adjacent to the television industry—networks, stations, advertisers, manufacturers, the FCC—the least examined is the trade association. Early forms of trade associations date back to the nineteenth century, and broadcasters looked to this cooperative strategy to help them fight royalty payments to the American Society of Composers and Publishers (ASCAP) in 1923. When businesses with an industry or trade in common want to direct the trajectory of that industry, they band together.[50] The ethos of the trade association is built upon the perception of threats from without, so the members turn inward and mobilize. Specifically, trade associations engage in a number of activities designed to benefit the functioning of all their members: "standardization of products, reduction of litigation, interchange of credit information, compilation of production statistics, stimulation of demand for particular materials or commodities, [and] encouragement of favorable legislation."[51] Activities like product standardization and information-sharing have led to accusations that trade associations impede competition among their members. Trade associations thus have a history of wrestling with antitrust legislation, so their motives, actions, and means all factor into their legality.[52]

In his overview of the association, Mackey points to the factors beyond the NAB's membership that signaled its vitality.[53] He mentions specifically the NAB's "friendships, contacts, and [. . .] the many thousand subsidiary persons and groups whose activities are coincident with broadcasting in all its phases."[54] These are the factors that Streeter calls the "historically embedded ensemble of social relations that make possible the production, distribution, and 'consumption' of the majority of commercial American television programs in the United States."[55] Because of these social relations, the NAB was able to, in Mackey's estimation, create "unity" and "stability."[56] And while Mackey argues that the NAB was "officially concerned" with both the creative and industrial sides of broadcasting, he concedes that the association focused first on regulation and finances, which its leadership believed enabled creativity to flourish.[57] Moreover, because the NAB needed to avoid charges of anticompetitive behavior and manage a multitude of opposing interests, both public and private, it had to operate under as much "calculated anonymity" as possible.[58] This reputation for working behind the scenes and avoiding the spotlight meant that the NAB did not distance itself from the larger stereotype of the secretive and self-interested trade association. It behaved as any other association would have behaved, despite the particularities of its trade.

The continued existence of trade associations underscores the viability of a regulatory territory situated in between individual businesses and government and in between the "extremes" of "unregulated competition" and "authoritative control."[59] More than simply being a happy medium, trade associations are, according to Streeter, one pillar of corporate liberalism, "a set of values" and practices that rests on the belief that government, corporate, and public interests can all be satisfied through their interaction with each other.[60] Although interaction is central to corporate liberalism, self-interest clearly propels the formation and preservation of trade associations. Commercial broadcasters' desire to regulate themselves stems, in a way, from their desire *not* to interact with outsiders.

For Harvey Jassem, broadcasters' early regulations "were designed to attend to the concerns of the public, the regulators, and the advertisers, while—or so that—broadcasters could go about their business in the least fettered and most protected fashion."[61] The basic goal of self-regulation, Jassem argues, is "to enhance or protect their profit position."[62] That protection is engendered by the industry's activities and, according to some scholars, by the government's inactivity. For example, Gaye Tuchman argues that the FCC's "symbolic regulation" is responsible for the industry's economically fruitful self-regulation.[63] Other structural reasons for this type of self-regulation are the common financial motives of networks and their affiliates.[64] However, that close relationship has marginalized independent stations, whose particular programming and advertising concerns sometimes put them at odds with the NAB. As appealing as a trade association might sound to individual businesses bound together by a shared trade, not all stations thought membership was worth the cost.

A persuasive letter from an NAB board member to a station manager in 1948 explains the benefits of belonging to the NAB.[65] The station manager—Milton L. Greenebaum—asked why he should pay dues to the association (and adhere to the NAB's advertising time restrictions) when other stations benefited without the expense of membership. The board member—Harry Bannister—replied that the NAB essentially operated as a crisis management firm, fighting incessantly against multiple foes, including journalists, religious organizations, intellectuals (or the "so-called intelligentsia," in Bannister's words), women's organizations, the government, music publishers, and the recording industry in general. In a speech in 1950, the NAB's director of government relations, Ralph Hardy, pled the same case in somewhat more dramatic fashion.[66] Without a unified and strong association, Hardy said, the industry would be defeated easily by the foes that Bannister mentioned. He warned his audience, "I have seen the campfires of the enemy; I have smelled the smoke from their forges where they are fashioning new tools with which to decimate and reduce to impotency the powerful giants of communication which are entrusted to our management." Industrial apocalypse apparently awaited a splintered broadcasting community. Behind the fearmongering were practical, if superficial, defensive tactics designed to assuage outsiders. One such tactic was the creation of standards to regulate content.

Censorship or Regulation?

The structure of broadcasting, fashioned by a coming together of public and private bodies committed to a commercial paradigm but tied to the murky idea of the public interest, left broadcast content vulnerable to restrictions that took the form of either regulation or censorship. As Hendershot explains, the FCC levying a fine at a broadcaster that airs a profane word is an example of regulation, a legal action. A judicial determination that the FCC overstepped the bounds of its authority in levying that fine means that the FCC did not regulate: it performed an illegal action by censoring.[67] Hendershot points out that although the two actions are "technically different," they are "often indistinguishable" from each other.[68] Tillinghast directs us to some foundational reasons for this—reasons written into key communications legislation.

As I discuss at greater length in chapter 5, the public interest clause in the Radio Act of 1927 is a slippery one that affected both the FCC's understanding of its authority and broadcasters' understanding of their own responsibilities. As we begin to consider content in a post-1927 context, we must remember that, as Tillinghast argues, the public interest "standard"—the baseline articulation of "guidance" and expectations for the administrative agency and the yardstick by which the agency's actions are evaluated by "reviewing bodies"—may have been incompatible with one part of both the 1927 act and the Communications Act of 1934.[69] Section 29 of the Radio Act of 1927 reads:

> Nothing in this Act shall be understood or construed to give the licensing authority the power of censorship over radio communications or signals transmitted by any radio station, and no regulation or condition shall be promulgated or fixed by the licensing authority which shall interfere with the right of free speech by means of radio communications. No person within the jurisdiction of the United States shall utter any obscene, indecent, or profane language by means of radio communication.[70]

Section 326 of the Communications Act of 1934 reiterates this statement specifically for the newly created FCC. The obvious contradiction in this passage immediately forecloses the possibility of free speech on the airwaves. Exceptions to the protections afforded by the first part of the Radio Act's Section 29 pepper broadcast history, and some are discussed in this book. Furthermore, Section 29's promise of freedom had two structural deficiencies that Tillinghast highlights. First, the censorship prohibition had to be included because the First Amendment's applicability to broadcasting was not automatic. Second, Congress had the power to revoke Section 29, creating a "fragility" that prevented it from "be[ing] equivalent to a constitutional guarantee."[71]

A third important check on speech was the license. David Silverman states that broadcasting enjoys limited protection under the First Amendment precisely because of the government's power to issue licenses. These licenses, Silverman further argues, are "issued for and held accountable to the public good or, more specifically, ill-defined community standards—a constantly changing and highly variable target."[72] In my study of the precarious position of astrology on TV in the early 1950s, I showed that both the public interest standard and concerns about fraud compromised the government's fidelity to Section 326 of the Communications Act in its license renewal hearings.[73] Not coincidentally, in its own standards the NAB echoed the FRC's and then the FCC's negativity toward astrological programming. This shared attitude had tangible, if inconsistent, effects on license holders, television astrologers, and devotees of astrology.[74] Surely, that shared attitude also blurred the distinction between public and private content regulation and between regulation and censorship.

Codes, Codes, and More Codes: The NAB and Early Radio

As private bodies free from the constitutional implications of content regulation, the NAB and the networks were able to monitor radio programs and impose standards of decency. For example, in "Broadcast Content Regulation and Cultural Limits, 1920–1962," Murray examines specific instances in which the broadcasting industry policed race, gender, and sexuality, and he uses a wide lens to explore both the deployment of power and the expressions opposed to that deployment. Using case studies like Mae West's sexually suggestive "Adam and Eve" routine on NBC's *The Chase and Sanborn Hour*, he connects the regulation of content to normative representations that maintain "imbalances of power."[75]

Murray's critical analysis of the ideology of content regulation contrasts with the type of noncritical institutionalist research that seeks to uncover patterns of behavior in the industry, but both offer crucial insights. Unpreoccupied with the ideological underpinnings of industrial maneuvers, the explications of the pre-1948 radio codes offered by both Morgan and Mackey nevertheless construct a foundational understanding of the ways in which the industry's self-regulation almost became a ritual—a routinized, redundant exercise focused on appeasing the government and other critics.[76] This research fixes our attention on the problems of standard-setting within the context of radio-era public-private relations.

The NAB adopted its first radio code, the "Code of Ethics," in 1928. Comprising six points, this code stressed the reach, service orientation, and positive aspirations of radio.[77] The association deemed this abstract code to be weak and quickly began revising it. Upon reflection, the NAB realized that its code needed to perform two functions: deflect criticisms and create unity among industry participants.[78] The 1929 code was more pointed than its predecessor, addressing offensive material and statements, fraud, unhealthy products, and deceptive advertising. The second part of the code established standards for industry behavior. This code discarded the abstract ideals of the 1928 code, and it included a clause about investigating violations. However, no investigations ever transpired.[79] As Morgan puts it, the 1929 code "lapsed into an obscurity from which it was not to emerge for several years."[80]

The Roosevelt administration's attempt to regulate self-regulation via the National Industrial Recovery Act (NIRA) Code in 1933 set up a Code Authority that applied to all stations and not just to those run by NAB members, as the previous codes had done. After the NIRA was deemed unconstitutional in 1935, the NAB had no viable self-regulatory document until it approved a "ten-point Code" in 1935.[81] This new code combined the major concerns of the two parts of the 1929 code, and accordingly, broadcasters mostly ignored it. Mackey argues that broadcasters' internal and external commercial struggles, plus a general exhaustion with rules, contributed to the snub.[82]

The NAB finally drafted a stronger radio code after noting the FCC's mounting impatience for network programming and practices. In 1938, a series of controversial programs, including the shockingly realistic sci-fi drama *War of the Worlds* on CBS and Mae West's "Adam and Eve" routine on NBC, triggered a public scolding by FCC chairman Frank McNinch. In that same year, the FCC launched an investigation into the networks' monopolistic practices, and RCA president David Sarnoff subsequently endorsed self-regulation during the FCC's inquiry. According to Mackey, Sarnoff's call "was immediately picked up by the NAB."[83] The drafting of the 1939 code yielded the strongest code to date, but broadcasters were unimpressed and reacted negatively.[84] The committee charged with crafting this code buckled under the weight of the broadcasters' distaste for the document—a document that had no real way to punish violators even though it called for the formation of a Code Compliance Committee. The revised draft avoided "specific provisions" and instead opted for "general platitudes" regarding broadcasting standards.[85] The NAB gutted its original version and settled for a brief document covering children's programs, controversial issues, educational programs, news, religious programs, and commercial dealings.[86] No mention of "ensuring compliance" appeared in the final draft.[87] Five-sixths of the NAB membership voted in favor of the reworked 1939 code, which Morgan describes as "a puny attempt to strengthen self-regulation."[88] Puny or not, Mackey argues that it did not spring from a vacuum: the NAB had in the back of its collective mind a serious government inquiry that eventually led to the sale of one of RCA's two NBC networks.[89]

The 1939 code maintained the appearance of self-regulation but in practice bore little resemblance to the detailed standards that would follow in the next decade. Despite its superficiality, the controversial issues clause in this code successfully stymied the notoriously vitriolic paid programs of Father Coughlin. Mackey explains that Coughlin went off the air because stations voluntarily complying with the code refused to do business with the demagogic radio personality, and not because the NAB took steps to enforce its provision that "time for the presentation of controversial issues shall not be sold."[90] The provision that squeezed Coughlin off the air came under negative FCC scrutiny in 1945 and ultimately required revision. Consequently, the NAB fashioned a 1945 code that functioned as an even weaker set of guidelines concerning "public questions, news, children's programs, education, and religion."[91]

Morgan attributes the subsequent radio code to the 1946 publication of the FCC's *Public Service Responsibility of Broadcast Licensees*, also known as the

"Blue Book." The Blue Book's influence on the NAB, and certainly on the 1948 radio code, is another example of the interconnectedness of government regulation and industrial self-regulation. Because the Blue Book "represent[ed] one of the most progressive initiatives by one of the most progressive Federal Communications Commissions in US history," according to Victor Pickard, it inflamed the leadership of the NAB.[92] The Blue Book concerned itself with radio stations' programming and license renewals. Specifically, it addressed the FCC's need to study stations' programs and their connection to the public interest before renewing a license. The Blue Book also printed the details of specific stations' license renewal hearings as case studies. Although the Blue Book noted the absence of a concrete policy guiding public interest–related licensing decisions, it explicated four ways in which broadcasters could program in the public interest: by supplementing commercial programming with sustaining programs (programs with no commercial sponsorship); by programming "local live" shows; by programming public issue–oriented shows; and by controlling "advertising excesses."[93]

The reaction to the Blue Book was swift, negative, and far-reaching. In a series of letters reprinted in an NAB pamphlet, then-NAB president Justin Miller explained to Representative Harris Ellsworth (R-OR) that the FCC had stepped outside the "proper scope of its authority" by wielding "undelegated power."[94] The "most dangerous" part of the Blue Book, for Miller, was its "statements of philosophy destructively inconsistent with a free medium of speech."[95] Furthermore, by showcasing several case studies of poor programming, Miller argued, the FCC painted all other broadcasters in a bad light. Miller also took issue with the way in which the Blue Book connected station wealth to quality of program service. In drawing attention to station wealth, the FCC implied that stations were doing so well that they could afford to improve programming; Miller's response was to accuse the commission of attacking stations' "freedom of speech."[96] In an earlier statement, Miller characterized the Blue Book as "a reversion to that type of government control and regulation from which our forefathers struggled to escape."[97] He continued in the subsequent letter to fabricate a link between regulation for the sake of controlling radio interference and regulation to control content, pinpointing the part of the Communications Act of 1934—Section 326—that denied the FCC the power to censor.[98] Miller's twenty-page letter (as reprinted in the NAB pamphlet), complete with footnotes, eviscerated the Blue Book and indicted the FCC's attempt to clarify as much as possible the unsettled connotations of the phrase "the public interest."

The American Civil Liberties Union (ACLU) disagreed with Miller. In a published response to broadcasters' opposition to the Blue Book, the ACLU called the FCC's actions "the most promising move for expanding the service of radio to the public."[99] The group also rooted the FCC's right to set public interest standards in the commission's licensing mandate. On the issue of freedom of speech, the ACLU argued that the FCC's recommendations, which focused on the "character" of programs and not their "contents," served the proliferation of speech rather than its reduction.[100]

Eventually Congress weighed in on the Blue Book. A House select committee investigated the FCC in 1948, concentrating primarily on the FCC's licensing and renewal policies, its Blue Book–related procedures, and its possibly unconstitutional behavior.[101] By the end of 1948, the House select committee had called off the investigation. The committee chairman, Forest A. Harness (R-IN), lost his reelection bid in early November, and upon resumption of the hearings in mid-November, the new chair, J. Percy Priest (D-TN), decided that the best body to continue the investigation was the Communications Subcommittee of the House Interstate and Foreign Commerce Committee.[102] The select committee did release its report, however. The committee recommended to the 81st Congress that the FCC kill the Blue Book and also chastised the FCC for hiring Charles Arthur Siepmann, a former BBC employee, to help draft the Blue Book.[103] "Such employment," the report read, "was a deliberate step toward government control of radio." FCC chairman Wayne Coy responded in a speech at Yale University:

> If freedom of radio means that a licensee is entitled to do as he pleases
> without regard to the interests of the general public, then it may reasonably
> be contended that restraints on that freedom constitute acts of censorship.
> If, however, the freedom of radio means that radio should be available as a
> medium of freedom of expression for the general public, then it is obvious
> enough that restraints on the licensee which are designed to insure the pres-
> ervation of that freedom are not acts of censorship.[104]

Pickard writes that the Blue Book had virtually no policy impact and that its impotence (and death) represented a win for self-regulation.[105] I argue, however, that the Blue Book made a sizable impact on the relationship between the NAB and FCC and on the rhetoric of the NAB in advance of the adoption of the Television Code (which will be discussed in chapter 3). That the FCC would presume to position the public interest as an attainable goal of commercial broadcasting—*and to outline steps for broadcasters to realize it*—exceeded the NAB's tolerance for the regulatory agency. Additionally, the very public circulation of broadcaster missteps could only discredit the association that was supposed to have a regulatory grip on its membership.

Falling back into its routine, the NAB established another committee to draft a revised code in 1946. Various ruptures between independent stations and network affiliates, including arguments over limitations on commercial time, stalled the code's adoption and ultimately resulted in revisions. Morgan writes that the once-"stringent" code became "completely suggestive in tone" and lost its method of enforcement.[106] Still, the 1948 *Standards of Practice for American Broadcasters* was not light on detail, and it was closer in appearance to the eventual TV code than the previous codes were. A "Broadcaster's Creed" preceded the programming standards and outlined the radio industry's commitment to democracy, public service, economic health, and the general enrichment of all listeners.[107] This version even included dos and don'ts for different genres of programs, including public affairs, political broadcasts, children's programs, and crime and mystery programs.[108] Special attention was paid to sound effects, clearly as a result of the deceptively realistic *War of the Worlds* broadcast.[109] Despite its substantive improvements over previous codes, the loss of the enforcement mechanism rendered it impotent. As one model for the Television Code, however, the 1948 *Standards of Practice for American Broadcasters* symbolized a pattern of industry behavior that would persist into the television age.[110]

Conclusion

Reviewing key moments in the development of radio regulation allows us to see how broadcast content became a growing problem for governmental and industrial authority. Did radio content have First Amendment protections if the government's licensing authority approved and had the capacity to expel speakers? Did the FRC confine its attention strictly to character? And can we distinguish character from content, as the ACLU believed was possible?

As a nongovernmental regulator, the NAB faced no constitutional impediment in repeatedly setting standards that tried to rein in radio content. Mackey and Morgan stress that broadcasters ignored those standards, so it is difficult to claim that the NAB believed in the efficacy of its measures. However, the mere presence of the radio codes did speak to a larger cognizance of both the public reputation of radio and the possibility that the government could use its licensing power to adopt a more interventionist stance if it so chose. As we will see in chapter 3, the NAB used the government's attention to content as a way to lambast the entirety of government involvement in broadcasting. Indeed, the series of radio codes proffered by the association over three decades signaled that the government had the power not just to police content but to influence the NAB's behavior.

Early attempts at self-regulation were a series of negotiations and frustrating concessions. In his study of the ideological maneuvers behind broadcast regulation, Murray asks a fundamental question that connects the code rituals described in this chapter to the NAB's eventual approach to TV: "Why was the NARTB so slow to adopt a television code?"[111] He points to the "heterogeneity of interests" colliding and yielding a "disunity of purpose," but I argue that after the NAB welcomed television into its self-regulatory regime in 1951, the association's purpose was far too speedy and unified.[112] Looking backward, we can trace that unification to the NAB's history with the FRC and FCC and to the association's participation in the systematic marginalization of noncommercial radio broadcasters.

Significantly, as we look forward to the emerging television industry, we see that the NAB hesitated when faced with the prospect of overseeing a new technology. In fact, before the NAB broadened its purview, the radio-dominated organization had to contend with a smaller group that had the foresight to organize early members of the TV industry. The story of that upstart trade association is the subject of the next chapter.

CHAPTER 2

Distinguishing Television from Radio via the Trade Association

The Rise and Fall of the Television Broadcasters Association

Television histories do not address the National Association of Broadcasters' difficulties with television, so the historiography would leave one believing that television had been a fixture of the association since its development, or at least since the beginning of full-fledged commercial operations.[1] In fact, the NAB paid scant attention to television throughout the 1940s. In a sense, television was put on hold for the duration of the US involvement in World War II, and manufacturing and most broadcasting did stall. While CBS left television broadcasting from late November 1942 to early May 1944, and DuMont went dark briefly beginning in mid-1943, NBC stayed on the air in a limited fashion throughout the war.[2] The industry did not stop from 1941 to 1945; its interrelated components continued to plan, debate, and formulate. That planning drove demand for a trade association to steer and represent industry participants, from the most powerful players to the basic units of broadcasting: the local stations. The result was the Television Broadcasters Association—an organization deserving of its own place in broadcast history.

This chapter takes issue with the tendency to treat the NAB as the inevitable home of the US television industry. Put very simply, the TBA wanted television and the NAB did not—at least in the 1940s. As an association organized to facilitate the success of radio, the NAB boasted a mature infrastructure and a sizable AM membership. The TBA was an upstart, a small group of industry elites aspiring to treat television as inherently special and superior to radio. With only five stations in existence at the time of its founding, the TBA lacked a view from the bottom and, as a result, focused on representing the networks and manufacturers. Its challenge was convincing new stations to join when it lacked the station-based management the NAB had spent decades crafting. Who had television, why they had it, and why that changed makes for a story—long ignored—that highlights the social and institutional entanglements within the web of industry relations. The story also magnifies the tensions between the old and the new. Television was poised to surpass radio in terms of investment, profits, and cultural influence. While the TBA-NAB conflict exposes the radio industry's hesitation during that moment of transition, it also highlights what the NAB had done correctly. The NAB had integrated the personnel and agendas of radio stations into its structure and governance, and though stations' participation could be volatile, its absence would ensure failure.

A Trade Association by Television and for Television

Official credit for inspiring television's first trade association goes to Klaus Landsberg, an innovator on the technical and production side of broadcasting.[3] Hired by Paramount Pictures in 1941 to launch its Hollywood television station, Landsberg reportedly embarked on a junket in 1943 to garner "support for an organization which would represent the television interests."[4] According to a TBA publication, "television interests were disorganized and seemingly getting nowhere" at this time.[5] Landsberg was president of the Society of Television Engineers (STE), and that organization formally created the TBA in January 1944 at STE's Chicago convention.[6] Landsberg's role in the TBA's formation is illustrative of Paramount's multipronged approach to television investment in the late 1930s and 1940s. As explained by Christopher Anderson, Hollywood studios looked to television's potential to mitigate some of the industrial turmoil already under way.[7] Paramount's partial ownership of DuMont Laboratories was emblematic of such a strategy; as Michele Hilmes argues, the DuMont deal was not a random act of diversification. It coincided with the 1938 antitrust suit against the Hollywood studios, so the purchase, for Hilmes, "indicate[d] the direction at least one studio was preparing to take in the event of divestiture."[8] Paramount's link to a television trade association through Landsberg and DuMont—an original member of the TBA—was further evidence of the studio's comprehensive outlook for the industry.

Although we should remember Paramount as a player in the TBA story, we should not overstate the studio's involvement. An internal NBC memo written by O. B. Hanson, chief engineer and one of the original directors of the TBA, reveals that the Radio Corporation of America (RCA) Television Committee was actually responsible for creating the association.[9] The Television Committee approached RCA chairman David Sarnoff, and Sarnoff

signed off on the plans, but as Hanson wrote, "for obvious reasons the organization of TBA [. . .] came from a source on the West Coast." Hanson
went on to explain that some of the TBA's seed money, necessary for the
association to take action on television standards and frequency allocation
hearings in Washington, came from NBC's three memberships. These three
memberships—for NBC's New York, Washington, and Chicago owned-and-
operated stations (O&Os)—entitled NBC (owned by RCA) to three votes.
The "obvious reasons" for the obfuscation of the TBA's roots probably referred
to the need to distance the association from any one company. Landsberg represented Paramount and, by extension, DuMont, but he spearheaded the formation of the TBA in his capacity as president of the STE.[10] The TBA could
not maintain the appearance of an industrywide association if its origins could
be traced to David Sarnoff's office.

With only ten names on its membership roster, the TBA adopted the
slogan "Uniting All People" and illustrated that mission in its seal, which was
composed of symbols representing "the rural, the urban, scholastic and industrial."[11] The TBA was formed to celebrate television's ability to reach everyone
and, more importantly, to secure the interests of the television industry by
learning from the mistakes of early radio.[12] To that end, one of its initial goals
was to clear any obstacles that stood in the way of nationwide commercial
television broadcasting.[13] TBA members were acutely aware of the speed with
which television would grow once the war ended, so the association "plan[ned]
aggressive action" with regard to channel allocation.[14] Its first president was
Allen B. DuMont (followed quickly by J. R. Poppele of WOR-TV, an independent station in New York), and it counted among its early members NBC,
CBS, DuMont Laboratories, General Electric, Howard Hughes Productions,
and Television Productions Inc. of Hollywood.[15] As the TBA matured, it
formed committees to focus on each of the following aspects of broadcasting:
advertising, station operation, public relations, programming, education, and
membership. Much like the NAB, the TBA sent its members a weekly bulletin
with industry updates.[16] The TBA also convened monthly to remain abreast
of industry developments and, similar to the NAB, held regular conferences.[17]

TBA rhetoric relied heavily on its uniqueness as a consequence of television's newness. One TBA publication describes the television broadcaster
as a pioneer "with no precedents to follow—except those that might have
been set for related industries—and many of these do not apply."[18] Without
mentioning radio by name, the TBA distinguished television as "new ground"
that required new knowledge wholly distinct from the earlier medium. Because the association formed so early, TBA president J. R. Poppele argued in
one 1947 report, "no precedents had been set to hamper its free growth and

expression. Starting from scratch, as such, we are better able to guide our destinies than if we were hamstrung with antiquated precedents and 'musts.'"[19] Poppele's words were not as concerned with radio as they were concerned with the NAB. Its refusal to see the NAB's past as a model for its own operations left the TBA free to pursue the interests of television without giving much thought to radio. It was also free to avoid adopting the NAB's questionable policies, such as the multiple radio codes that sought to police radio content. Considering the strong ties its members had to radio, and understanding how connected radio was to television in matters of regulation and advertising, one could argue that Poppele's statements amounted to empty rhetoric. However, as the years passed, and as the TBA needed to attract more members to stave off encroachment by the NAB, Poppele's words transformed from the bluster of a young upstart into the rallying cry of a hardworking yet endangered institution. In 1950, Poppele vigorously defended the TBA's track record and value in the face of low membership numbers:

> TBA's fine record of accomplishments since it was founded in 1944 should commend itself to all broadcasters who are not now affiliated with the Association. TBA's greatest asset has been *its ability to speak without qualification for television broadcasters only*. Since it has been so vocal—and has done its job so well, despite financial limitations—it *deserves the unqualified support of the industry*.[20]

Indeed, the TBA accomplished much for the early television industry with little recognition and support.

As early as August 1944, the TBA focused on influencing the outcomes of the FCC's frequency allocation hearings, drafting a set of "allocations principles" that aligned specific channel assignments with proper television service in the public interest.[21] The FCC's acceptance of TBA involvement prompted the association to describe itself as the "official voice of the television industry in the United States."[22] The TBA's primary activities in 1945 revolved around allocations. Not only was the TBA a fixture at FCC hearings, but its Engineering Committee chief also directed the FCC's Special Industry Committee on TV Allocations.[23]

Intent on securing exclusive spectrum space for television, the TBA needed to participate in the debate over frequency modulation (FM) radio allocations.[24] Although the TBA sided with FM interests, a move that protected TV's stake in the very high frequency (VHF), the TBA was not opposed to ultra high frequency (UHF) television allocations, as RCA and DuMont were.[25] In 1949 (early in the license freeze), the TBA's Engineering Committee

proposed "full utilization of UHF" while the FCC reconsidered TV allocations.[26] This decision may have brought the TBA into conflict with RCA, a
broadcasting company that held all television manufacturing patents in VHF.
This conflict may have exacerbated tensions between the TBA and the NAB.
Conflict with the NAB was already present as early as 1946, when the TBA
successfully fought the NAB's proposal to revisit FM allocations and claim
channels already allocated to TV.[27]

   In addition to lobbying for channel allocations, the TBA directed its attention to the health of manufacturers. By early 1947, the association counted
among its members a number of "leading manufacturers of television equipment," a constituency affected by postwar labor unrest and attendant material
shortages.[28] Reflecting on the TBA's progress in 1946, President Poppele noted
that the number of television sets manufactured fell short of earlier predic-
tions because of these shortages and because the industry was "starting from
scratch" with "no moulds or dies or previous know-how to make a resumption
of manufacture a relatively simple procedure."[29]

   Also impeding manufacturers' progress was a problem relating to television
antenna installation in apartment buildings in New York.[30] Antennas clut-
tered rooftops and interfered with picture quality, so by March 1947 landlords
refused to allow antenna installation until a "master system" was created.[31] As
Lynn Spigel notes, DuMont and RCA began marketing television sets to consumers in 1946, signaling a transition from the public exhibition of television
to its private, home-based consumption.[32] Eager "to tap into and promote the
demand for luxuries," television manufacturers needed not just homeowners
but apartment renters, particularly in New York City, to spend their money
on sets.[33] Recognizing that the landlords' ban affected potentially two million
families and, consequently, dealt a "serious blow to manufacturers and servic-
ing companies affiliated with TBA," the association created its Subcommittee on Apartment House Television Installations and submitted to the Real
Estate Board of New York a plan to resolve the problem while manufacturers
designed a master antenna system.[34] By December 1947, the TBA Engineering
Committee had approved two systems, and one was put to use immediately.[35]

   Noteworthy, too, was the TBA's success in staving off a 20 percent amusement tax, or "Cabaret Tax," on television sets used in public establishments.
Implementation of the tax would have hampered the sale of sets to bars, for
example—a possibility that drove some sponsors to "withhold contracts for
televising big league games" until the matter was settled.[36] Expediently aban-
doning the rhetoric of TV's uniqueness, Poppele convinced Joseph Nunan of
the Internal Revenue Service (IRS) that television approximated radio more
than it resembled "cabaret entertainment."[37] Poppele also reminded Nunan

that television's public service role complicated its classification as "strictly an 'amusement.'" The IRS concurred and exempted TV from the tax.[38]

By the end of 1947, Poppele was convinced that the TBA had risen in "stature" because of its "initiative and progressive actions."[39] Evidence shows that the TBA championed television relentlessly in both major and seemingly minor ways. In 1947, the TBA convinced the *New York Times* to carry television listings on Sundays.[40] In 1948, the association officially protested the rates that AT&T and Western Union were charging the networks for interconnection.[41] And in 1950, the TBA successfully testified against the Treasury Department's proposed 10 percent excise tax on TV sets.[42] Examples like these demonstrate the TBA's will to reflect "the majority view of the industry."[43] Nevertheless, the organization's documented decisiveness did not hold up when the conversation turned to television content.

Anxiety about program content—apparent even before the widespread launch of mainstream commercial television—factored into the TBA's early plans for television. This type of anxiety followed the tradition of "moral panic" that accompanied the rise of "commercial amusements" in the early twentieth century. Spigel notes that reformers clamored for "codes of decency" in motion pictures and their exhibition venues.[44] Whereas entertainment in the public sphere triggered concerns about content, broadcasting as a "mechanized amusement" for the private sphere inspired "hopes for salvation."[45] The cultural sentiment Spigel describes coincided with changing conceptualizations of quality in broadcasting. According to Sewell, early discourse about radio and television quality centered on "institutional form, practice, and structure" rather than on "conventional conceptions of content."[46] Therefore, the theme of morality was less prevalent in the 1930s than in the 1940s and '50s. The TBA's conversations about content as early as 1944, then, marked a telling transitional moment in which form and content commingled as notions of decency began to creep into the preoccupations of the early television industry.[47]

The TBA drafted a detailed television code in July 1945, but never ratified it.[48] In 1946, the Program Committee worked on revising the initial code without formulating concrete plans for its adoption and enforcement.[49] The association's attitude toward such a code centered on "moral responsibilities," but its appreciation for television's infancy tempered any zeal to mandate standards. Poppele wrote in 1947 about shielding the medium's "vigorous experimentation" from "namby pamby do's and don'ts."[50] In the same report, however, Poppele encouraged TBA members to endorse "the idea of adopting some form of guide or code [. . .] until such time as a more permanent code can be adopted." Poppele wanted the TBA to embrace a sense of responsibility while avoiding the complications of a code—complications made evident by the repeated failures of the NAB to enforce its radio codes.[51] Noran E. Kersta, a TBA Code Committee member and manager of the Television Department at NBC, even stated that any code "accepted by the TBA Directors would never have unanimous acceptance or approval by the TBA membership."[52] Nevertheless, the Code Committee sought out "all of the various public informational and entertainment media codes" in existence to help them determine the proper course of action.[53] Still without its own code in 1948, and hesitant to stifle experimentation with "sight and sound," the TBA crafted "general principles of programming service" and distributed the 1948 NAB "Standards of Practice" and the Motion Picture Production Code for use by stations.[54] This decision (or indecision) conformed to Poppele's belief that a "guide" should precede a concrete code.[55] TBA correspondence indicates that the association knew a code was a losing proposition, especially so early in television's development. Even support for a code much "simpler" than the 1948 NAB code echoed the sentiment that experimentation needed to be free from the type of censorship that could result from moral panic.[56]

In hesitating to adopt an industrywide code, the TBA supported strong self-regulation at the level of the station; its never-ratified 1945 code emphasized that television "should be jealously guarded by those who should be best qualified to protect it—The Television Broadcasters themselves."[57] The TBA used its relationship with its station membership to differentiate itself from the NAB. TBA Code Committee chair Lawrence Lowman indicated such a motive in a letter in which he wrote that adopting the NAB code—which was his preference—"wouldn't be the best thing for the TBA to do."[58] And while the TBA Code Committee's statement of principles and policy proposed using both the NAB code and the Motion Picture Production Code as guides, the committee stressed that "individual" broadcasters would decide which standards worked best for them on a case-by-case basis.[59] The Code Committee went so far as to suggest that the broadcasters should do this to foster "public confidence and good-will," which were the "only foundation on which to perpetuate the democratic, competitive system of television broadcasting."[60] In skipping over the foundational Communications Act of 1934, the legislation compelling broadcasters to serve the public interest, the committee discursively erased the legal mechanism connecting television programs to license approval and renewal. But by not inserting itself as the arbiter of standards, the TBA deviated from the NAB's tactics and opted to support the ability of local television broadcasters to regulate their businesses. In fact, an early draft of the statement of principles asserted that the TBA "fully supports and subscribes" to the NAB's 1948 code.[61] This wholehearted support did not appear in the final draft circulated to television broadcasters.

The TBA represented itself as a television-only club despite some of its members' involvement in radio, and it had the makings of a strong association but for its existence in the shadow of the older, larger NAB. As it mimicked the NAB—though certainly on a diminished scale—it operated for four years outside of the NAB's competitive gaze. The TBA's most pressing problem was its small membership, a necessary evil because of the few television stations in existence. Stations, rather than networks, stood to benefit the most from both routine and extraordinary trade association actions. Radio stations in particular stood to lose if television interests competed with their own, but for radio stations moving into television service, the reality of two trade associations was less than ideal.

The NAB Considers Television

Sniffing around the Set: 1947–1948

In 1931, David Sarnoff predicted that television would have a "beneficial" impact on the radio industry, but he took for granted the enthusiasm with which the institutions in place to represent radio broadcasters would embrace television.[62] Sarnoff prognosticated later, in 1947, that radio broadcasters who had not ventured into television would see their radio revenue erode following the increase in television viewership.[63] His words eventually reached receptive ears. The NAB examined its stake in television at its annual conference in May 1948. Conference participants expressed "faith that sound broadcasting can live safely and profitably side-by-side with television."[64] Both Frank Stanton of CBS and Lewis Allen Weiss of the Don Lee Broadcasting System were optimistic about the new media landscape, but noted that AM broadcasting would bear the financial brunt of television during the new medium's formative years.[65] The NAB determined that 235 stations that represented only 18 percent of its AM radio membership but were responsible for 53 percent of the money the NAB received from its member stations were moving swiftly to television.[66] Fifteen of those radio stations operated television stations, 54 had construction permits to build television stations, and 166 were awaiting word from the FCC regarding their license applications.

The NAB was finally talking about TV, but it appeared to be moving too slowly for Walter J. Damm, vice president of WTMJ in Milwaukee, Wisconsin, an NBC affiliate. In June 1948, Damm sent a letter to all television stations in operation, conveying a message of frustration.[67] Having run a television station since December 1947, Damm was convinced that neither the NAB nor the

TBA was doing much for station operators, who suffered from an ignorance of each other's policies, rates, and contracts. The NAB simply was "not concerning itself with television," and the TBA was "doing nothing except what it can on a budget of around $20,000 per year." The formation of the National Television Film Council just five days prior prompted Damm to ask, "How many more of these associations are we going to have?" His call to action was simple. Station managers needed to get answers from the NAB, or from the TBA if the NAB was unwilling to reply. Damm's ultimate solution was for the two associations to merge in order to create a mechanism by which TV stations could share information and standardize their operations.

Damm had experience with just such a merger. In the mid-1940s, he had been president of FM Broadcasters Incorporated (FMBI). FMBI, which launched in 1940, "merged" with the NAB's FM branch in 1946, but members did not see special benefits from this enterprise until 1948.[68] Damm drew a parallel between that merger and the trouble he foresaw with bringing television into the fold. In a letter to NAB president Justin Miller notifying him of his missive to all television stations just two weeks before, Damm wrote:

> For my part, I would suggest a completely separate section of the NAB, separately financed and staffed and completely undominated by a board of AM broadcasters. This will no doubt be looked upon with considerable disfavor. However, FM's experience has proven the point which I made at the time FMBI and the NAB merged. You cannot expect an AM broadcasters dominated board to look with favor upon spending the money put up by AM broadcasters to service FM or television.[69]

Television and radio needed to be "under one tent," in Damm's estimation, but they also needed to be segregated to circumvent the AM domination Damm feared. Pushing for a speedy resolution, Damm also recommended a name change if the NAB decided to adopt TV.

The stations' responses to Damm's initial letter (at least the responses he forwarded to Miller) expressed doubt about the TBA's efficacy, owing mostly to its financial constraints, and opted for an NAB-led enterprise. Miller was appreciative of Damm's efforts and agreed that the NAB needed to move on the television issue.[70] As to Damm's doubts about AM broadcasters' feelings, Miller wrote, "Even the AM operators are becoming interested in television so rapidly that I think we will have little difficulty in getting action at the next meeting of the Board." Miller anticipated three possible courses of action: initiate activity by the NAB board, amend the NAB bylaws to incorporate television into the structure of the NAB, and pursue a "coordination" with the TBA as they had with FMBI. "I am doubtful as to whether this would be feasible," Miller wrote, "as there seems to be considerable pride of prestige in the present TBA organization." This was not the last time someone from the NAB would remark on the TBA leadership's attitude. Damm shared Miller's pessimism about a merger.[71] Nevertheless, Miller and Damm agreed to set up an August meeting wherein television station managers could meet and discuss how to move television forward within a trade association framework.

Sizing Each Other Up: 1948

In anticipation of the August 11, 1948, meeting of television stations, Miller fashioned a board of directors' Television Advisory Committee (TAC), which would determine how best to fold television into the NAB.[72] But first the television stations needed to meet. In attendance were representatives from all five radio and television networks and twenty-three television stations. TBA director George Burbach spoke with mixed feelings.[73] After covering the highlights of the TBA's brief history and emphasizing the association's successes despite its underdog status, Burbach aired his skepticism about the feelings and actions of the NAB's AM membership. He felt that AM stations would hesitate to share his "deep conviction" that television was "the greatest public service media ever devised," surpassing even radio. More importantly, a dominant AM membership would have to learn to share the umbrella association, which "would have to be prepared to go 'all out'" for television. Burbach concluded by endorsing cooperation between the TBA and the NAB, while Frank Russell of NBC backed a "federation"—an amalgamation of associations representing related segments of the industry—that "would assure a unified front against common enemies." Summing up the August meeting, the NAB's press release touted the "industry progress and great economies" that attendees agreed cooperation between the TBA and the NAB would bring.[74] The two groups would proceed by forming two three-person committees to investigate such cooperation.

The NAB's Television Advisory Committee met in Chicago just two days later to discuss the role of television within the NAB. The results of that meeting, at which Harry Bannister was elected chair, ensconced television within the protective bounds of the NAB, if informally. The TAC argued that the creation of a "complete television trade association service" was the NAB's "absolute duty."[75] The rationale for this new development grew out of calculations about radio's future. Because radio broadcasters would fund television's launch, the NAB needed to commit itself to the new pursuit of its primary membership. Ostensibly, the NAB would not forget the TBA as it forged ahead; the TAC promised to work with the television association "to achieve unity." The TAC also put forth two resolutions to solidify television's place within the NAB. The first would put television members on equal footing with radio broadcasters and ensure television representation on the board of directors. The second, echoing the resolution from the August 11 meeting, would form two three-person committees to determine how the NAB and the TBA would move forward together.

The TBA and NAB committees met on September 1, 1948.[76] A. D. Willard Jr., vice president of the NAB and a member of the NAB's three-person committee, reported back to NAB president Miller after the meeting and laid out two options: cooperate or compete.[77] He doubted the viability of cooperation, describing the TBA delegation as "extremely sensitive people who are inclined [. . .] to place personal matters ahead of industry welfare." The NAB was an entrenched trade association meeting with a much younger and much smaller organization. That Willard detected some defensiveness on the part of the TBA's leadership may have been a misinterpretation of a considerable power imbalance.

The NAB-TBA plan that emerged from the committees' follow-up meeting in October detailed administrative, membership, and financial structures.[78] The plan indicated that the TBA would stop being a full-service enterprise and instead would take charge of "representation of television interests in the controversial fields of TV promotion, TV allocation and TV legal representation." The NAB would simply extend to television what it had already performed for its radio membership: "regular trade association service in non-controversial areas in the fields of sales promotion, advertising, research, labor relations, public relations, programming, [and] legal (where it is non-controversial)." One upside other than the legitimation of television within the NAB would be the windfall from new TV members—at minimum a total boost of $25,000 to $30,000 to the NAB-TBA coffers.

The TBA was not necessarily banking on a merger. In between the September meeting and this one, the TBA board of directors requested that TBA secretary-treasurer Will Baltin prepare a five-year budget forecast for an expanded TBA—a TBA, in other words, that would assume all the functions for television broadcasters that the NAB did for its membership.[79] Anticipating that 75 percent of television licensees would join the TBA, Baltin calculated an annual income of $105,000. A full-service operation including legal, engineering, research, labor, public relations, and special services departments would cost the association $102,000 annually. Baltin sent his report to the TBA board after the NAB-TBA plan was drawn up. Money would be tight, but the small association's projected expansion was feasible.

Less than one month after the formulation of the NAB-TBA plan, the NAB board directed a new committee to conduct a big-picture study of NAB operations.[80] A "realignment" was proposed to confront "the problems in all fields of electronic mass communication in order to provide adequate representation and service to all such interests."[81] TV's and FM's increasing momentum triggered the study, but the consequence of this new resolution was the postponement of two major TV-related developments: the creation of the NAB's Television Department and the NAB-TBA merger. "The TBA cooperation project is not out the window," *Broadcasting* magazine stated, but it had stalled indefinitely.[82]

The TBA held its annual clinic in December, and after J. R. Poppele was elected to his fifth term as president, he spoke about the TBA-NAB merger plans. Evincing the "pride" and "sensitivity" that the NAB leadership had commented upon negatively, Poppele stated, "Your directors are of the firm conviction that TBA must never lose its autonomy and that your industry problems can best be handled in an atmosphere where television—and only television—is the object of one's particular interests."[83] He then revealed that the TBA-NAB merger plan was just about to be approved by the TBA board when the NAB announced its realignment study. Poppele positioned the TBA as acting in the "best interest of the television industry" when the NAB abruptly aborted the plan.[84] Significantly, the TBA board vote would not have been unanimous: Allen B. DuMont (who *Broadcasting* aptly noted had "no alliance with sound broadcasting") opposed the merger on the grounds that "the television broadcasters and manufacturers [could] put their best foot forward only through an organization exclusively devoted to the espousal of the visual cause."[85] The year 1948 ended with the TBA digging its heels into an imagined landscape dotted only with television sets. Perhaps trying to appease the majority of its members, the NAB waited.

Going Their Separate Ways: 1949

The TBA exploited the NAB's hesitation deftly, implementing changes that attempted to bring its services on par with the NAB's operations. In February 1949, the TBA announced the addition of three new member services.[86] These services included a "monthly program exchange service to provide a complete list of new programs on member stations," "quarterly 'status of the industry' reports," and monthly reports of regulatory and legislative actions. The TBA also planned to update members regularly on license and construction permit applications as well as station personnel. Finally, the TBA was moving toward the fulfillment of Walter J. Damm's earlier request: the formation of committees to handle issues such as copyright and advertising rates. Two months later,

Poppele continued to emphasize TBA excellence at the NAB convention. In a statement to attendees, Poppele celebrated television's accelerated growth. Noting inevitable complications for the television industry, Poppele remarked that the TBA "from the outset has recognized these problems and today is in a far better position to wrestle with them than ever before, by virtue of its dominance in television. By cooperative action with other groups interested in the welfare of the industry, the TBA *will do the job.*"[87]

Two developments in May continued to pit the TBA against the NAB's entrenchment. First, the NAB board voted to install a director for its TV Department.[88] Recall that the NAB postponed the creation of that department during its realignment study; although the NAB apparently followed through on the TV Department, it did not restart merger talks. As the NAB board decided on a director, the TBA openly courted FCC chairman Wayne Coy for the presidency of the TBA—a newly created, salaried position based in Washington and tailored to improving public and governmental relations.[89] The TBA consistently described its relationship with the FCC positively, so its pursuit of Coy is unsurprising.[90] Its willingness to spend the money was surprising. The TBA struggled financially, so DuMont was to raise the necessary funds for the three-year appointment by appealing to television manufacturers.[91] *Broadcasting* speculated that this sort of expansion would incite the NAB to pursue television "aggressively."[92] At the time, the NAB counted only six television stations among its 1,838 members—three more TV stations and 81 fewer radio stations than in 1948.[93]

In June 1949, *Sponsor* magazine—an advertising trade publication—pointed out the flaws in the NAB's treatment of television. Calling for a federated NAB, the trade magazine likened the NAB's "absent" treatment of television to that of FM radio.[94] Using AM station dues to fund a competitor was simply too unpalatable to the NAB membership. "The result," *Sponsor* wrote, "is an association (Television Broadcasters Association) which month by month is becoming a more important factor in the visual air advertising field." Citing the failure of the NAB to incorporate the TBA, the article lamented the "divided" state of broadcasting, which had three associations: one for AM radio, one for FM radio (the FM Association, FMA), and one for television. This situation was simply bad for business, as it fostered division when these media had interests in common.[95] The article proposed that AM, FM, and TV each have a board of directors with separate dues structures, and that all operate under a "Federated President."[96] Though "internal politics" would be inevitable, *Sponsor* believed that the status quo was too shortsighted in the face of technological developments.

Broadcasting's own Research Department polled broadcasters and found that just over 50 percent favored a plan to reorganize the NAB into three

departments—one each for FM, AM, and TV.[97] Just over 24 percent of broadcasters preferred a merger with the TBA. NAB president Miller acknowledged these statistics and the "many pressures" placed on the NAB "to get into
television" in a letter to a station executive just two days after *Broadcasting*'s
poll was published.[98]

By August 1949, the NAB had begun to envision itself as an "all-industry
association" with audio and video divisions serviced by all of the association's
departments.[99] As the NAB reorganized its structure, it abandoned the idea
of a merger with the TBA.[100] Crucially, with only six television members,
the NAB could not create an autonomous television branch. Instead, Miller
planned to treat TV like FM—"as a service organization and with propor-
tionate representation on the NAB board."[101] The TBA would continue to be
the organization that promoted TV—much as the FM Association handled
FM broadcasting—but the NAB would not engage in promotion since it
was dominated by AM broadcasters. Signs in August that both the FMA
and TBA were ramping up their operations did not seem to interfere with
the NAB's plans for a "comprehensive" FM and TV operation.[102] The NAB's
ambitions were bolstered by news in late August that NBC's O&Os were set
to join the new TV division.[103] In fact, the NAB was ramping up its own TV
operation with a dramatic increase from six to thirty-two TV members (of
seventy-eight TV stations in operation) in September.[104] The NAB managed
to lure the bulk of these members, which were affiliated with existing AM
members, by discounting their dues ($10 a month) until January 1, 1950.[105] The
association was not just luring members away from the TBA; it also poached
the TBA's vice president, G. Emerson Markham, for the position of Video
Division director in early September.[106]

Strife inside the NAB: 1949–1950

Markham's departure from the TBA colored the NAB's behind-the-scenes
membership troubles in a personal way. Network participation in the NAB
had not been steady, and radio station membership fluctuated too. In June
1949, several large radio stations—all NBC affiliates—left the NAB, which led
Miller to hypothesize that these particular stations needed to "cut corners" to
accommodate their large investments in television.[107] Stations were not the
only businesses blaming their departure on financial difficulties. In a letter to
Miller, ABC's Mark Woods also mentioned "expense" as one reason the network was looking to sever its ties to the NAB; ABC's most recent bill from
the NAB was $5,000.[108] In reply to Woods, Miller articulated his realization

that network happiness depended on television station "expansion," and that the NAB had a major role to play in facilitating that. In other words, one way to keep the networks on the membership roster was to be an advocate for television.

In February 1950, Miller felt sufficiently threatened by the possibility of the networks resigning from the NAB and taking their O&Os with them that he wrote to the NAB board of directors to ask "how far the NAB should go in an effort to perform services of particular value to the networks and to their owned stations."[109] Money again was an issue. The combined network and O&O contributions amounted to $7,731 per month, which Miller pointed out was just a little under the amount contributed by 625 radio stations. Such a financial hit would disturb the very structure of the NAB. Miller also worried that the networks' affiliates might follow suit. The majority of responses to Miller's query assessed that the cost of losing the networks was too great to deny them special services.[110] Other board members relayed their exasperation. Citing the networks' power to "get practically anything they want" and the need to treat all members equally—from the largest network to the smallest independent station—these dissenters did not want to be held hostage by elite bodies interested only in services pertinent to themselves.[111] One board member, Harry R. Spence, posed the prescient question: "Are they satisfied with the NAB television department?"[112]

Miller believed that NBC would stick with the NAB, but he was less confident about ABC and CBS.[113] Robert Kintner, president of ABC, was a TBA board member and, according to Miller, "probably resent[ed] a little [the NAB] pulling George Markham away from TBA." Both Kintner and Frank Stanton, president of CBS, were concerned about money, and Joseph Ream, executive vice president of CBS, had a number of reasons to leave the NAB. Miller felt that these reasons were exaggerated, however, and stated that CBS simply did not "want a strong NAB." CBS resigned from the NAB on May 18, 1950, and ABC followed on May 26. Both took their O&Os with them (CBS had seven, and ABC had five). The NAB stood to lose $40,000 per year because of CBS's departure, as well as between $25,000 and $30,000 from ABC's resignation.[114] Although CBS officially attributed its resignation to redundancies across the network's and the NAB's functions, Miller dismissed these excuses and called the networks "selfish," characterizing them as ungrateful opportunists who were "passing on to their affiliates [. . .] the burden of maintaining their trade association just because they think there are no immediate pressing issues."[115] Miller also faulted Kintner for being too sensitive, since he felt that the NAB's new "Standards of Practice" "unjustly classified ABC's programs as below standard."[116]

This NAB-network turmoil was indicative of the general upheaval surrounding TV. In 1949, television was on the verge of occupying a central place in US social and economic life. The 1949 revenues for the television industry amounted to just over $34 million, "almost four times the 1948 amount," according to the FCC.[117] And in late 1949, the NAB became aware that the 1950 US census would include a question about television set ownership.[118] The signs of television's import proliferated, but the NAB had yet to move television beyond the committee level. The NAB Television Committee, chaired by Robert D. Swezey of ABC affiliate WDSU-TV in New Orleans, operated strictly in an advisory capacity, meeting biannually to review the processes and policies of the NAB as they pertained to television.[119]

Toward the National Association of Radio and Television Broadcasters

Before the NAB decided to promote television within its organization, the TBA, emboldened by its new ten-point "pledge to the industry," launched a massive membership campaign in June 1950.[120] The pledge consisted of ten goals for 1950, including ending the license freeze, monitoring broadcasters' programming "responsibilities," and opposing unfair government intervention.[121] The new ten-point program represented a financial commitment to the industry's well-being that required more support from station operators than it currently had.[122] In 1950, the TBA still counted only 35 stations—about one-third of television broadcasters in operation—among its dues-paying members; thus, a handful of television stations were subsidizing work that supported all television stations.[123] By contrast, approximately 1,500 dues-paying radio members of the NAB subsidized the work that benefited its 37 TV members.[124] Faced with this disparity, the TBA planned to make personal appeals to all nonmember stations.[125] The campaign was a public relations push of sizable proportions, and it did not go unnoticed.

At the fall 1950 meeting of the NAB's Television Committee, item number two on the agenda was the "place of the TV department in the Association," indicating that television station operators were not going to be content simply with an advisory committee.[126] One station president, Campbell Arnoux of NBC affiliate WTAR in Norfolk, Virginia, wrote to Miller in early November 1950 to campaign for the creation of an "autonomous [television] group with its own sub-board of directors, its own dues schedule and committee."[127] Chiding Miller for the NAB's "inadequate" service, Arnoux called for immediate change lest the NAB "lose the TV stations to the TBA."[128]

By mid-November 1950, the NAB board of directors decided to act on the television resolutions adopted at various NAB district meetings earlier in the fall.[129] Robert D. Swezey, chairman of the TV Committee, submitted the committee's proposal to the NAB board and added this justification: "There is purpose now, as television stations move into the profit columns, in establishing a sound economic basis for their continued membership in the NAB and the definition of a procedure which will give television members necessary independence and flexibility of action."[130] The NAB board resolved to establish "a separate TV Board of Directors [. . .] to operate within the framework of the NAB, but to have complete autonomous authority to determine its own dues structure and to formulate its own policies with respect to all questions relating to television."[131] The resolution called for the NAB's TV members to convene to sort out the details, and it created yet another committee to bring the recommendations of the upcoming convention to the NAB board.[132]

One day after the NAB board adopted this resolution, the NAB sent telegrams to all of the TV stations in operation—not just the NAB members—notifying them of the development and assuring them that its adoption would entitle TV members to all of the NAB's services.[133] The NAB estimated that a revised dues structure that included new and existing TV memberships would net the association anywhere from $89,000 to $114,000 per year.[134] The NAB's priorities were clear: expand its service to television while protecting its radio membership.

The NAB later reported that it had received "enthusiastic" responses to its news.[135] Approximately forty television stations that were already members of the NAB approved of the formation of NAB-TV, and the rest of the nation's television stations "show[ed] interest" in the venture.[136] Significantly, the NAB also mentioned that many TBA members not already a part of the NAB were "favorably inclined toward the new NAB-TV."[137] The NAB also expressed "no [. . .] intention of interfering with the TBA or absorbing it," but it did pronounce its superior ability to function as a legitimate trade association for television.[138]

However, a possible TBA-NAB merger was not forgotten. As late as December 11, 1950, a TBA committee awaited a meeting with the NAB to discuss folding the TBA into the newly formed NAB-TV.[139] The TBA's president, J. R. Poppele, conveyed an eagerness "to assure TV trade association autonomy for [the] best interests of [the] industry."[140] Paul Raibourn of KTLA in Los Angeles chaired the TBA committee responsible for negotiating with the NAB over the creation of one television trade association.[141] Merger talk was just that, however, and nothing ever materialized. Even though the TBA had decided to attend the January NAB-TV meeting, Raibourn admitted that the new body "would mean the demise" of television's first trade association.[142]

If everything went according to plan, the NAB expected NAB-TV to be fully operational by April 1, 1951.[143] Surpassing the purely advisory role of the original Television Committee, the new Television Board would have five primary duties, including the power "to enact, amend and promulgate Standards of Practice or Codes and to establish such methods to secure observance thereof."[144] No such code existed yet, but the NAB surely was looking ahead to mounting concerns about the content of television programs.

Swezey openly doubted the ability of the TBA to compete with what the NAB was proposing. The TBA lacked both the membership and the infrastructure to launch a completely new television trade association on par with the NAB's plan. Doing so would, according to Swezey, "require a minimum investment of several hundred thousand dollars" if the new association were to match the strength of the NAB.[145] While some television operators preferred to see the TBA and the NAB merge, both Swezey and NAB Television Board member Harold Hough stressed that there would be no such merger.[146]

At a January 1951 meeting held by the NAB, 75 percent of TV stations in attendance pledged their "unanimous and enthusiastic support" for the new body, which would, in effect, require that everyone "abandon the TBA."[147] An editorial in *Broadcasting*, a publication exceedingly friendly to the NAB, described the atmosphere as one of "almost unprecedented harmony."[148] Members displayed overwhelming, if not unanimous, support, voting "by a 15-to-1 ratio" for the new NAB structure.[149] An additional marker of success was the high number of stations attending that had aligned with neither the NAB nor the TBA.[150] Paul Raibourn, representing the TBA, warned that "autonomy cannot exist unless each group under the NAB is its own court of last resort."[151] Assured that the blueprint for NAB-TV would secure that degree of autonomy, Raibourn announced that the TBA would soon "wind up its affairs."[152]

Following the Chicago meeting, the NAB rewrote its bylaws to reflect television's new status within the association. The new bylaws created a strong and autonomous Television Board with complete control over all television-related matters, including policies, dues structures, and meetings. The new Television Board would be "its 'own court of last resort.'"[153] In a showy bit of rebranding, and recalling Damm's suggestion in 1948, the NAB changed its name "to be more descriptive and to give television and radio equal recognition."[154] The NAB officially became the National Association of Radio and Television Broadcasters (NARTB) on April 1, 1951.[155] The TBA officially dissolved on the same day.[156]

Big changes ensued, but the results nevertheless reflected television's minority status. The new NARTB structure consisted of three boards of directors with the TV board holding a maximum of fourteen members (compared to the radio board's maximum of twenty-five).[157] Elected members would fill

FIGURE 2.1. "You and your family are in this picture!" (NARTB pamphlet, 1956)
This NARTB pamphlet explains the importance of viewer feedback to the
proper functioning of the association. Courtesy of Wisconsin Historical Society,
WHS-133142.

nine seats—one of the seats was filled by Paul Raibourn of the TBA—and
representatives from each of the four television networks would fill the re-
mainder of the vacancies.[158] Television stations without radio holdings had
to have a minimum of two seats.[159] Miller vacated the presidency in June 1951
and assumed chairmanship of the board, leaving a hole filled by Harold E.
Fellows.[160] The NARTB anticipated that sixty TV stations would join immediately. Each of the networks was entitled to one membership; the NAB
dropouts ABC and CBS were predicted to return to the fold.[161] The manu-
facturer membership of the TBA, meanwhile, was "urged" to add their names
to the NARTB roster once the TBA dissolved.[162]

Why Television? Why Then?

A study of the New York metropolitan area conducted by NBC assessed the
growth of television from 1950 to 1951, a period in which circumstances had
"drastically changed," according to NBC.[163] Indeed, more than 10 million
homes had television sets by January 1951, compared with 200,000 three years
earlier.[164] Two million of those television homes (serviced by seven stations)
were located in the sixteen counties studied by NBC.[165] Among the findings
that confirmed the NAB's decision to incorporate television into its association were: TV families made $644 more per year than non-TV families;
TV families bought 73.2 percent of new cars purchased within the previous

six months; TV families were larger and had younger children; the heads of TV families were younger than heads of non-TV families; more TV families owned telephones, refrigerators, and cars than non-TV families; and TV owners spent 135 minutes of their day watching television, versus 61 minutes listening to radio, 47 minutes reading newspapers, and 11 minutes reading magazines.[166] Even non-TV owners enjoyed viewing television: the study showed that 59 percent of non-owners watched TV in a month, and 45 percent of non-owners watched TV in a week.[167] When the study combined owners and non-owners, it found that 73 percent of family heads, regardless of set ownership, watched TV weekly.[168] These statistics underscored the urgency of the two trade associations' situations.

When we insert the TBA's story into the larger history of television, the collision of old and new media replays through the lens of the trade association—a lens that brings into sharp focus the inseparability of industrial and social relations, the division between businesses that adapt and those that lag, and the centrality of broadcast stations to the development of the television industry. The young association was poised to represent the nation's multiplying television stations and had been working for the benefit of television since 1944. Unwilling to alienate its AM radio membership, the NAB glanced at television now and then, failing to commit to a lasting relationship until AM stations moving into TV pushed the issue.

The details of the TBA's fall and the NAB's internal struggles realign TV history with the basic unit of broadcasting—the station. As integral as the networks were to the NAB's financial health, local broadcasters filled leadership positions in the association and influenced the group's agenda. The TBA's inability to attract a majority of television stations revealed that stations did not occupy the same political space in the TBA as they did in the NAB. NAB documents show, however, that fear of losing network members forced them to admit that the bigger companies were worth special favors. Regardless, the future of the TBA—as a stand-alone association or as a partner with the NAB—hinged on AM broadcasters. As much as the TBA wanted a space dedicated only to TV, the reality of the broadcasting industry bound TV to radio, and the reality of trade associations bound the TBA to the will of stations.

In this chapter, I have focused on how the NAB and the TBA defined themselves in relation to their media and their membership. In the following chapter, I examine how the two associations approached television content and how medium specificity factored into the matter of self-regulation.

CHAPTER 3

The Industry Talks about a Television Code

Discourses of Decency, Self-Regulation,
and Medium Specificity

A pre-TV NAB publication describes its association as a "voluntary organization working to promote and defend, when challenged, the American System of privately owned, competitively operated radio. A self-governing organization comprising an overwhelming majority of American broadcasters whereby, through rational self-regulation, they give their answer to the cry for Government operation."[1] Formed in 1923, the NAB grew in power as the radio industry did.[2] And as is evident from the first two lines of its mission statement, the association defined itself in overtly political and economic terms. Its enterprise was private, and it existed to protect US commercial broadcasting from a coercive government.

That coercive government first took the form of the US Department of Commerce, then the Federal Radio Commission (FRC) in 1927, and finally the Federal Communications Commission (FCC) in 1934. The regulatory agencies changed, but the NAB remained intact to monitor signs of government overreach, which, in the NAB's assessment, frequently appeared in the agencies' interpretations of the public interest. The broadcasters' obligation to operate in the "public interest, convenience, and necessity" originated with the Radio Act of 1927, but it carried over to the Communications Act of 1934 without much clarification or expansion. "The public interest" is, as Patricia Aufderheide observes, "a phrase of art, and deliberately a loose one."[3] The absence of one concrete definition resulted in multiple definitions colliding over everything from UHF television channel allocations to news editorializing. This is a well-documented story, but examining it more closely reveals an almost untenable situation. The literature stresses that Congress neglected to insert specific guidelines because it was unwilling to figure out how to regulate new technologies.[4] In other words, Congress passed the problem along to the FCC, which had to contend with a mobilized industry outwardly suspicious of government.

The NAB had supported and regulated the commercial radio industry since the early 1920s and the commercial television industry since 1951, when it changed its name to NARTB—the National Association of Radio and Television Broadcasters. Part of its burden had been the creative and cultural dimension of its central trade. In 1951, the NARTB felt threatened enough to create a document—the Television Code—that, among other things, restricted television content. The NARTB's push to regulate television programs and advertising was not out of character for the trade association. As discussed in chapter 1, the NAB drafted multiple radio codes from the late 1920s through the late 1940s, but could not legally enforce them. In retrospect, an NARTB television code was probably inevitable, but at the time of its inception, such a move was controversial and involved many sensitive moving parts.

This chapter explores how the early television industry conceptualized content in three different contexts. Since the Television Broadcasters Association—TV's first trade association—tackled television before the NAB did, I examine how the smaller organization discursively tied content to decency. While some in the TBA embraced run-of-the-mill pro-decency rhetoric, others warned of the dangers of constraining content too early in the development of the medium. In chapter 2, I argued that the TBA struggled partly because it catered to manufacturers instead of cultivating a strong relationship with local broadcasters. As a beleaguered trade association inherently linked to radio interests but dedicated to facilitating television's growth, the TBA struggled to argue for television's—and television content's—uniqueness within a regulatory framework constructed according to the specificities of radio.

Following my discussion of the TBA is an analysis of the discourse of the dominant trade association. At the center of this section is the NAB's discourse of self-regulation leading up to the implementation of the 1952 Television Code, a document that attempted to answer television's vocal critics by constructing a paradigm for wholesome content. The NAB's strategy was to valorize the television industry's work as the exercise of democracy and free expression. In the process, the NAB demonized any government entity that attempted to direct television content in the public interest. Pitting private industry against a caricature of totalitarianism at the outset of the Cold War, the NAB sought to equate commerce with democracy, thereby naturalizing commercial broadcasting as a democratic, fully American institution.

To tie both trade associations to the process of regulating content, the final section explores how they and other organizations navigated around medium specificity. Was television analogous to other media that had been subject to industry codes? How would early ideas about TV's uniqueness hold up in the face of older self-regulatory models? In moving toward an industrywide code, different groups advanced competing ideas about the best course of action, but in actuality their conflicts stemmed from the extent to which they cared to see television as a wholly different medium.

The TBA and Industrial Philosophies of Television

A 1944 TBA news release touted the success of the organization's first conference, held on December 11 and 12 in New York City's Hotel Commodore.[5] Attended by representatives of the broadcasting industry, the motion picture industry, advertising agencies, publishing, equipment manufacturers, and the US government, the conference was expected to draw 1,000 for the evening banquet. According to the press release, Allen B. DuMont "expressed the hope that when the second Conference is held [. . .] peace will have been restored and the expansion of television as a national service shall have started."[6] All manufacture of television sets ceased after President Franklin D. Roosevelt declared a state of emergency in 1941. Those factories dedicated to television were refashioned to aid the war effort. In late 1941, nine television stations broadcast to approximately 22 million people.[7] Eighty license applications awaited FCC approval.[8] These conditions hung over the TBA conference as speakers pondered the ways in which television would develop after the war.

Among the topics discussed on the opening day of the conference were the pre- and postwar "structure" of television programs, "post-war television network facilities," the "important role of the television broadcaster," and the various roles in television for "program producers, advertising agencies, publishers, performers, theatres, manufacturers and broadcasters."[9] In a nod to the star of the gathering, entertainment for the first evening was not performed in person but televised on twenty-eight TV sets in the hotel ballroom by stations WNBT and WABD. First was a telecast of boxing matches, and after that was an RKO-produced film of the morning's conference proceedings.[10] Work preceded entertainment, of course. Panel discussions were broken up into seven television-related fields: advertising, manufacturing, producing, theater television, print publications, talent, and broadcasting. Allen B. DuMont opened the conference and prepared attendees for "a most comprehensive review" of the pursuit of "commercialized television."[11] DuMont stressed that commercial TV would only be achieved through coordination and "group action."[12] When J. R. Poppele took over for DuMont, he reported on the TBA's membership, which tallied thirty-three; he appeared to be proud of that number, given the TBA's brief time in existence.[13]

Conference speeches varied in topic and tone, but the issue of content surfaced again and again, for different reasons. The twin themes of "home" and "decency" frequently accompanied speakers' thoughts on programming. Dr. Walter R. G. Baker, the first chairman of the National Television Systems Committee, remarked that television would need to balance "innocence" and creativity.[14] The home's "standards of decency," "humor," and "code of conduct" had the potential to differentiate television from movies or the theater, so cognizance of viewers' expectations would keep television from becoming "an invasion."[15] Lee Cooley of the Ruthrauff & Ryan advertising agency focused exclusively on responsible programming. Although the engineers had given birth to television, Cooley remarked, it was up to the producers "to see that this infant art is guided along the path that will lead to its greatest development."[16] Tracing the implementation of codes in the motion picture and radio industries and underscoring government oversight of radio, Cooley anticipated more of the same for television, since it was "free" and would "invade" US homes.[17]

Although the destination of programs occupied some speakers' minds, the origins of programs also entered into the conversation about content. Walter S. Lemmon from International Business Machines (IBM) spoke of constructing relay systems that would move content *to and from* metropolitan areas and less populated areas. His point was that the audience should not be at the mercy of New York and Los Angeles for its programming; one way to correct for the more liberal value systems of the East and West Coasts would be to create two-way programming flows. "Certainly," he argued, "the cultural aspirations of America should not be judged entirely by the offerings of those particular localities. I hope television will encourage the development of programs in many other American communities and then 'import' them as well into the metropolis."[18] Lemmon spoke of culture, but the values or morality communicated by New York– and Los Angeles–based programming were another facet of the one-way, network-to-affiliate transmissions that rankled some in the TV audience, in the FCC, and in Congress. Ultimately, a model like Lemmon's would have conflicted with the tight control over production exercised first by advertising agencies and then by the networks.

The consensus on the television producers' panel was that "some sort of standard" needed to be established, but the panel's chairman believed that the advertising agency should not be the one to create that standard.[19] The participants did not resolve the problem of "questionable items" making their way onto television screens, but they did offer suggestions, such as hiring staff members to help censor live programs.[20] Predictably, the solutions rested in self-regulation rather than in government oversight.

The first to mention a code was John F. Royal, vice president of television for NBC. Royal stressed the value of an association such as the TBA when he called for "cooperative" and "collective" standard-setting.[21] He went so far as to propose that the TBA "give careful consideration to an industry programming code which will, as far as possible, guarantee the cooperation of all television advisors in presenting clean programs."[22] There could be "no compromise with decency."[23] This invocation of a code was noteworthy, given NBC's reticence to pursue that course of action later. And in 1944, Royal's idea seemed less like a first step for television and more like a continuation of the radio industry's self-regulation.

This inaugural TBA conference raised many questions, offered some solutions, and evinced a sense of optimism at the new beginning embodied by the young trade association. The ideas shared among the participants provide insight into the industry as it anticipated a postwar escalation in television service and demand. Having traversed many difficult roads with radio and emerging somewhat wiser, the established television interests may have felt that they were starting down a relatively smoother path. Realistically, however, the addition of sight to sound invited some complications that could be conquered with technological expertise and others that required political wrangling.

The second TBA conference was held two years later, on October 10 and 11, 1946, in the Waldorf Astoria Hotel in New York City.[24] By that time, the TBA had grown to just under fifty members and counting.[25] TBA president Poppele described the conference as "herald[ing] the start of television's national expansion," and the conspicuous presence of NBC cameras documenting the proceedings made this expansion apparent to the participants.[26]

Not skipping a beat, this conference revisited the sacred space of the home and television's place within it. Poppele's opening remarks lauded the TBA's role in helping the television industry to achieve "commercial expansion," but he quickly identified the "great challenge" facing that industry at the moment: "moral responsibility."[27] Echoing Royal's speech from 1944, Poppele stated, "There can be no compromise with decency."[28] Poppele characterized this responsibility as greater than any one broadcaster; he viewed it instead as an honorable pursuit worthy of the entire association's attention.

It has been apparent to the Directors of TBA during the past year, that if television is to succeed as the greatest means of mass communication yet conceived, and as a monumental contribution to public service, it must be clean and wholesome, completely tolerant, fair in all public issues and a welcome visitor into the American home.[29]

Poppele touched on the potential problems in multiple forms and genres—religion, public affairs, drama, and comedy—and pointed to the "peculiarities" of television that would require unique approaches to presentation.[30] He assured the audience that TBA members were "giving careful study to all material, so that a high standard of clean, wholesome programs may be maintained on all television stations in the country."[31]

Whereas Poppele and others at the 1946 conference spoke about decency as though it were an objective and benign value, another speaker dared to broach the consequences of mandating decency. In speaking on the topic of good taste, the Mutual Broadcasting System's (MBS's) Frank Kingdon warned about the sacrifices inherent in the pursuit of such standards. The intimacy of a television broadcast reaching into the home brought with it a "danger," according to Kingdon.

> The minute that you get an industry [. . .] that does touch people intimately and effectively, then all the receivers, all the conventional forces of our society rise up and try to check that instrumentality. [. . .] And all the societies for the suppression of vice, and most of the societies for the suppression of thinking will immediately get active, to make every effort they can to prevent television from experimentation. So I think the television programmer is going to find himself caught between a natural sensitivity to his social responsibility, and yet a desire to make his program interesting and even sensational.[32]

Kingdon blamed commercialism too, in a surprising turn, for the avoidance of experimentation and the prominence of "over-conventionalism" and formulaic storytelling.[33] So-called good taste needed to be policed by the public, but Kingdon cautioned his audience not to abandon experimentation, "even at the expense of outraging or disturbing certain sections of the community."[34] He concluded by demanding that the industry "pay the necessary costs of experimentation, so that television will come out with its own form of art, unfettered by the conventionalism and the precedents of its predecessors."[35] While the term "experimentation" was used at the conference mostly to signify stations' classifications (those not operating commercially but experimentally), Kingdon adopted it as a way of describing television's aesthetic and narrative forward motion.

Television's commercial forward motion persisted into 1948 and struck the roadblock known as the FCC's license freeze. In December of that year, the TBA held a clinic at the Waldorf Astoria Hotel in New York City with approximately 400 broadcasters in attendance.[36] By 1948, the number of television stations in the United States had increased from 17 to 48, the number of television sets in homes approached one million, and 312 applications awaited FCC approval.[37] The proceedings of this clinic were considerably briefer than those of the conferences, but several contributions by speakers are worth noting here. An important warning came from the clinic's keynote speaker, FCC chairman Wayne Coy. After applauding the audience for maintaining a sense of decency absent an established code, he urged them to do more to preserve television's "good name."[38]

> The American home is not a night club. [. . .] If you take precautions now not to be tempted to the primrose path, you will be saving this art from the excesses, the remorse, the clamor for reform, the struggles for redemption that plague, in varying degrees, almost every other form of communication. Television's good name is in your hands. To preserve it unsullied you will be called upon to be vigilant, unswerving, prepared to shut the gates of mercy upon the first offender.[39]

Inspired by Coy's words, Poppele suggested future industry action when he said, "I believe from your remarks will come the code that we feel should play a big part in the growth of television."[40] Although Poppele's admiration for Coy's line of thought reveals a problematic link between the private body and the government regulator—one that reemerges in chapter 5—its positive tone helps us to understand the TBA as a young organization trying to build something new while navigating old regulatory terrain. The TBA's failure to follow through on its own code later that year is further evidence of this agglomeration of prickly circumstances.

But Poppele's comments also remind us that the TBA was not averse to a code; it simply had not agreed on one yet. The TBA wedged its abandoned 1945 code in between Hollywood's production code and the NAB's radio code, selecting content standards specific to a visual medium while addressing the circumstances of commercial broadcasting. At a length of ten pages, the 1945 TBA Code was guided by (if not lifted entirely from) the Motion Picture Production Code. Consider, for example, its first general principle: "Correct standards of life, subject only to the requirements of drama and entertainment, shall be presented."[41] This is the second general principle of the Motion Picture Production Code. Vague as well as ideologically fraught, this first of three general principles—all from the Production Code—precedes five and a half pages of explicit prohibitions that also appear in the Production Code. In its final pages, the TBA Code transitions to broadcasting-specific rules. Standards relating to children's programs, controversial issues, educational programs, news, religious programs, and advertisements draw from and elaborate on the NAB's standards. Notably, this code resists the lengthy and unrestrained moralizing of the Motion Picture Production Code and opts instead for a pithy appeal to the public good in its four-sentence-long preamble.

That the TBA did not move forward with its 1945 code fed the belief that the industry had not held television content to a high standard—or to any standard for that matter. In an April 1950 *New York Times* opinion piece entitled "Anarchy in Television?," Edward Lamb, a newspaper and television station owner, bemoaned the lack of accountability in the industry.[42] The government could not censor, he wrote, so some stations neglected public interest programming altogether. Lacking such oversight, the industry could not be trusted to abide by its mandate to serve the public interest. Self-regulation should be the answer, but the industry, Lamb argued, had no trade association: "The Television Broadcasters Association has within its membership only a minority of the telecasters, and it is doing nothing whatsoever to act as spokesmen for the industry." The NAB was no better, in Lamb's estimation, since it was "torn asunder" by radio and television interests and made "no attempt to act on any of the growing pains facing the industry."

Poppele responded to Lamb in the following week's *New York Times*.[43] Exposing Lamb as a "fence-sitter" who sought to criticize without making the investment to join the membership rolls, Poppele stressed that the TBA had been active since its formation. He wrote, "Since TBA came into being, the official records of the Federal Communications Commission and the archives of television itself are literally jampacked with details relating the role this small but effective organization has played in setting the pattern for television's commercial introduction and expansion." As a result of the TBA's small membership (in 1950 its dues-paying members totaled 35), a handful of stations, he pointed out, were subsidizing work that affected all stations, Lamb's included. By contrast, the NAB's January 1950 TV membership, though numbering only 37, was subsidized by approximately 1,500 dues-paying radio members.[44] Frustrated by such a disparity, Poppele asked, "Would it not be better for all concerned if those broadcasters who bleat for greater self-regulation would avail themselves of the agency which exists and which recognizes the need?"[45]

Although Poppele protested any claims of the TBA's ineffectuality, his framing of the issue proved Lamb's larger point. Poppele highlighted the TBA's role in getting *commercial* television off the ground; Lamb was lobbying for more educational programming by nonprofits and referenced state-run public service channels in other countries. Lamb believed that the government should do more to advocate for the public, but beyond the nods to decent programming in the previous years' conferences, the TBA focused only on the commercial paradigm with limited government involvement. Lamb's assault on both the TBA and the NAB immediately preceded a sea change in the television industry's self-regulatory framework. The TBA was vulnerable, and the NAB knew it. By April 1951, the TBA was a non-issue.

Early conversations about television's presence in the home did not differ substantially from the conversations that took place well after commercial television's successful integration into daily life. Industry leaders allergic to government input championed decency on their own terms. But closer examination of the mid-1940s exchanges—those in which a philosophy of television began to take shape—makes apparent a relative uneasiness with the radio model of standard-setting. Formulaic, safe programming and industry-set standards were the norm, but why should they translate easily to television? As we saw in chapter 2 and as we see here, the TBA was anxious to differentiate itself from the NAB while remaining beholden to the same institutions that propped up radio. The fact that the TBA drafted a code but never implemented it exposes the tension at work within the TBA's mission. The association's reluctance to adopt a code was emblematic of its desire to do things differently than the NAB did. As I show in the next section, the NAB's mode of operation was entrenched in the commercial system and strategically incorporated content into its ultimate goal of naturalizing commercial broadcasting.

The NAB's Discourse of Self-Regulation

Eager to preserve the private, commercial backbone of television, the NAB sought to reconcile the creative dimension of its trade with the demands of an advertiser-supported, government-regulated medium. The NAB's discourses of self-regulation leading up to the implementation of the 1952 Television Code framed the conversation about television content as a dire struggle between the democratic principles held by capitalists and the tyrannical intentions of a coercive government. By studying the manner in which the NAB characterized its work privately and publicly during television's infancy, we can draw connections between the splintering of broadcast regulation, the interplay of regulatory cultures, and the resulting blueprints for television content.

Internal NAB discussions are fascinating displays of strategy, honesty, and sometimes animosity, arrogance, and victimization. Chapter 5 discusses these confidential missives in greater detail; here the focus is on the face the NAB showed to its peers and to the public. Clarke writes that trade associations can simplify the work of government regulators because, in the system known as "corporatism," the two bodies can reach agreements "semi-formally and informally."[46] This approach to regulation renders private any contentious bargaining and constructs for the public eye an industry that appears "reasonable, responsible, and cooperative."[47] The record shows that the opposite occurred in the early days of television regulation. Although NAB leadership and staff carried on perfectly amiable personal relationships with regulators and elected officials, the NAB's public speeches were rife with acrimony toward the government. By bracketing the analysis of the NAB's rhetoric within the context of two overlapping sets of circumstances—the launch of the Cold War and the postwar expansion of television—this section offers a sense of the atmosphere in which the NAB moved toward TV and confronted the matter of TV content. Juxtaposed with the previous section, this analysis also highlights stark differences between the NAB's tactics and the TBA's tactics.

The anti-Communism of the late 1940s and 1950s infected the television industry in multiple ways. One visible and impactful manifestation was the publication of *Red Channels: The Report of Communist Influence in Radio and Television* in 1950. As Thomas Doherty notes, the 151 entertainers named in *Red Channels* "ranged from lockstep Communist Party hacks, to mainstream liberals, to bewildered innocents."[48] The identification of so-called radicals was just one fragment of a much more expansive project. Susan Brinson discusses the ways in which the anti-Communist sentiments and activities of the 1940s and 1950s have been linked to two dominant historical narratives relevant to the regulation of broadcasting. Brinson acknowledges the theory that the Soviet Union actively tried to depose the US government, but she invites us to consider two more mainstream theories.[49] The first posits that the climate of anti-Communism was a conservative reaction to the progressive changes initiated by the New Deal. The second theory departs from the sociopolitical motivations for the "backlash" and engages instead with economic motivations.[50] For many conservatives, the New Deal represented the encroachment of "government into corporate affairs," so New Dealers were liable to be branded Communists because of their endorsement of "antibusiness" policies.[51] Brinson explains that the FCC was itself a target of conservatives because its anti-monopoly rules, formulated in the early 1940s under chair James Lawrence Fly, were in keeping with New Deal policies.[52] Recall from chapter 1 that Charles Siepmann's involvement in the drafting of the Blue Book agitated both the NAB and the House select committee tasked with investigating the FCC. The reasons for the Red Scare thus provide crucial context for the NAB's discursive maneuvers as a trade association devoted to commercial radio but looking ahead to commercial television.

The primary mouthpiece of the NAB at this time was Justin Miller, a former academic and judge, who assumed the presidency of the NAB in 1945. The strength of his leadership during such a transformative period in the history of broadcasting cannot be overstated. But his contributions to the reactionary political atmosphere in the late 1940s should be just as prominent in the NAB's history. A review of some of Miller's speeches can be mapped closely onto Brinson's framework for understanding the intersections of politics, economics, and broadcast regulation. A 1947 speech by Miller positioned the NAB as a canary in a coal mine, warning everyone of impending government censorship.[53] Evoking the country's forefathers and the freedoms they fought to secure, Miller connected everyday criticism of radio to a greater irresponsibility. When we criticize commercialism or genres like soap operas and demand that the government step in, he reasoned, we contribute to our own destruction. His appeal to self-regulation—"I hope the broadcasters of America will [. . .] eliminate all just causes of complaint and thus remove the temptation which incites to government control"—admitted some flaws in the private system, but he warned, mentioning Hitler and Mussolini by name, that the industry's failure to self-regulate would transform the FCC into a stateside Nazi propaganda machine.

By 1949, while hinting at the need for a "standards of practice" for television, which the NAB had not yet absorbed into its structure, Miller continued to rail against the government in his speeches. In April 1949, he attributed nothing less than "lawless[ness]" to all administrative agencies.[54] He further distanced the good intentions of private enterprise from the coercive strategies of government by encouraging all listeners to agree with "the proposition that it is government which has always threatened freedom; however much it may have disguised its actions and pretended they were for the common good." He continued on to argue that Marxists had propagated the myth of a broadcast industry monopoly—a myth that would lead to government control and, even worse, a system like the British Broadcasting Corporation (BBC).

Early in 1950, Miller stopped speaking in hypotheticals and began addressing actual complaints about television. That viewers were complaining about television in such numbers as to prompt reaction from the NAB contrasts sharply with the critical evaluation of the 1950s as the "golden age" of television. As Boddy notes, critical and industry assessments of television at the time were much more complicated than what popular memory recalls.[55] Hierarchies emerged that privileged intimate, realistic, sixty- or ninety-minute-long live dramas, but not all programming was of that ilk. Many different genres populated programming slates, including news, soap operas, sitcoms, variety shows, quiz shows, children's programs, and sports. Genres waxed and waned, as was the case with the decline of the live anthology drama and the

FIGURE 3.1. "Attacks on Freedom of Communication" (speech by Justin Miller, 1949) NAB president Justin Miller accused the government of illegal censorship in this 1949 speech. Courtesy of Wisconsin Historical Society, WHS-133411.

rise of the filmed western. As will be discussed in chapter 4, each genre, including cooking shows, had the potential to offend or anger viewers, and some in the audience did not hesitate to contact stations, networks, sponsors, and even the FCC to make their complaints known. Even in the earliest days of mainstream television service, viewer complaints about low necklines and profane language poured into the FCC and television stations around the country. The general manager of a radio station in Milwaukee wrote to Miller early in 1950 to assert that something needed to be done about television content. Miller agreed, but other factors may have been at play.[56] Talk of government subsidies for broadcasting, which would have led to greater regulation, appeared in several of Miller's letters, so that, too, may have inclined him toward a more organized approach to the television problem.[57]

In response to a *Christian Science Monitor* publication on children and television, Miller let loose a diatribe tinged with fear. He worried openly that government involvement in content would stunt the medium and exacerbate the FCC's paternalism. Referring to the government in all its forms as the "Great White Father," Miller's words were sounding increasingly desperate.[58] In spite of the almost hyperbolically critical face that Miller displayed in his speeches, he turned a friendly face toward FCC chairman Wayne Coy after Coy spoke at the NAB's 1950 convention in Chicago. In thanking him for attending and speaking, Miller shared credit for improvements in broadcasting with Coy.[59] Outwardly, the NAB claimed progress in the quality of broadcasting, but Miller's strongly worded speeches still painted the FCC as the oppressor and the NAB as the perpetual victim and crusader. The NAB employee who actually dealt with the government on a daily basis—Ralph Hardy, the director of government relations—managed to push Miller's rhetoric even further. He described his experiences testifying before Congress and the FCC as "frightening" and as a "baptism by fire."[60] The man whose job it was to bridge the gap between the government and the NAB viewed the regulators as a "hideous three-headed monster of prosecutor, judge, and jury." For the NAB, the walls protecting it from that monster could not be high enough. Ultimately, the crafting of the Television Code was evidence of panic—not panic about censorship, as the NARTB would have had everyone believe, but panic about a decentering of the commercial television model.

Inching toward a Television Code

As the TBA meetings discussed at the beginning of this chapter indicated, talk of a code was alive and well in the late 1940s. Justin Miller congratulated his organization for propagating its own codes since, in 1948, television and comic books looked to follow in the NAB's footsteps.[61] By then, the TBA had set up a Code Committee and, in lieu of prematurely setting standards, opted to circulate copies of its "Statement of Principles," the NAB's "Standards of Practice for Radio," and the Motion Picture Production Code to its member stations.[62] The TBA's "Statement of Principles" stressed "good taste," "fairness," "responsibilities," and the "public interest"—all buzzwords to demonstrate that the industry and the regulators shared similar values.[63] But as discussed in chapter 2, the TBA's statement never mentioned government regulators. Instead, it highlighted the autonomy of broadcasters. Each station would need to monitor itself using the documents the TBA supplied as a guide. This case-by-case approach to self-regulation embraced the tenets of localism rooted, for better or worse, in the dominant culture of each community and its own standards. More practically, however, this approach relieved the national organization of a tremendous burden. Drafting a code was a hurdle, but instituting an enforcement mechanism that would not run afoul of antitrust legislation was daunting.

At the network level the movement toward television-specific standards was not so slow. In 1948, NBC published *Responsibility: A Working Manual of NBC Program Policies*, which recycled its radio standards for both radio and television.[64] NBC's in-house censorship body, the Continuity Acceptance Department (CAD), had been in operation since 1934, and in 1948 it issued a bulletin at the close of one year's worth of television censorship.[65] The bulletin reported that both radio and television advertising and programs shared the problem of adhering to "good taste," but advertising on TV suffered from compromised "believability and honesty." In addition to referencing its work on advertising, the bulletin noted the CAD's handling of two separate suicide scenes in two different dramas and a "hypnosis demonstration" in another program. The CAD also emphasized its work previewing questionable visual gags, as well as Milton Berle's slapstick routines; the department even checked up on Berle's jokes via a "confidential advance outline" since Berle preferred to withhold major jokes from rehearsals.

As time passed and offended viewers conveyed their feelings to sponsors, stations, networks, and the government, grumblings about the television problem prompted some action. In February 1950, Miller asked Hardy and G. Emerson Markham (the NAB's Video Division director) to pursue standards for television.[66] Ominous statements by FCC chairman Coy about television programming constituted "fair warning" for Miller that the industry needed to examine the issue. Specifically, in reply to a question about television's regulatory problems, Coy had stated, "Right now that whole situation is in the

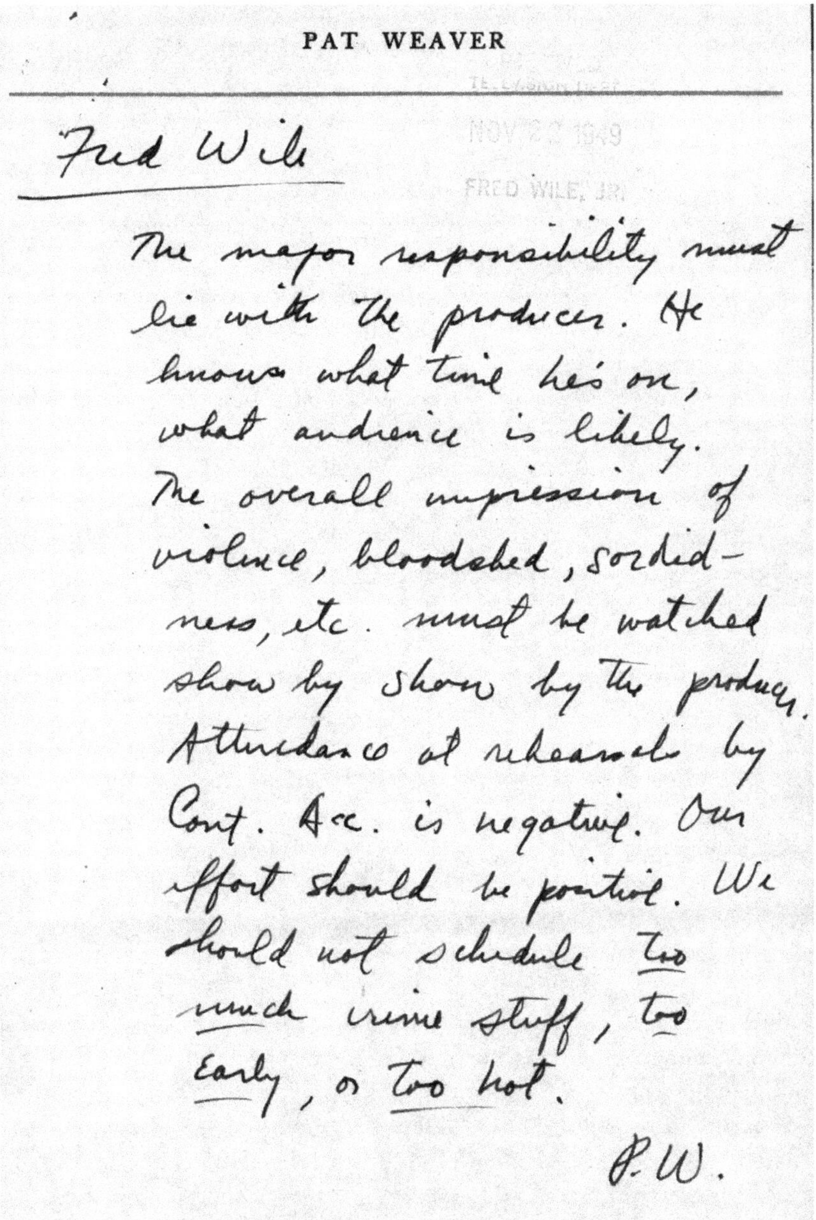

PAT WEAVER

NOV 22 1949

FRED WILE, JR.

Fred Wile

The major responsibility must lie with the producer. He knows what time he's on, what audience is likely. The overall impression of violence, bloodshed, sordidness, etc. must be watched show by show by the producer. Attendance at rehearsals by Cont. Acc. is negative. Our effort should be positive. We should not schedule too much crime stuff, too early, or too hot.

P.W.

FIGURE 3.2. Pat Weaver memo to Fred Wile (1949)
Pat Weaver's 1949 memo stresses the producer's responsibility for program content and refers to the presence of censors at rehearsals as "negative." Courtesy of Wisconsin Historical Society, WHS-133409. Image courtesy of NBCUniversal Media, LLC.

hands of the broadcasters. If they recognize the problem, the need for tighter regulation may be obviated. To the extent that TV may ignore the needs and feelings of the community, its problems may be serious."[67] With great haste, the NAB decided to survey all of its member stations, including the networks, about their programming policies. A letter from Markham to Pat Weaver, NBC's vice president in charge of television, gives an impression of the letter NAB sent to all of its members.[68] In his letter, Markham interpreted Coy's remarks to be a harbinger of government intervention. "While it may be too early to think now of a highly definitive industry code," he wrote, "it appears not too early to accumulate preliminary information on the manner and extent to which television program subject matter is being monitored." Markham wanted to know if NBC had its own set of policies, and Carleton D. Smith (NBC's director of network television operations) replied a few days later, enclosing NBC's 1948 *Responsibility* publication.[69] Smith assured Markham that NBC had applied its high radio standards to television, and in a form that was even more "stringent" than the radio code, but he warned against setting standards "which will affect adversely the growth of television." Exactly what he meant by that is unclear.

Markham was not deterred, however, and on March 27 he wrote to Miller expressing his "fervent desire" to implement standards for TV.[70] Internal politics complicated that desire and threatened to obstruct the NAB's plans. First, the networks did not want a code. They were responsible, not the stations, for distributing most of the programs that had riled up critics. Second, the NAB was not on the best terms with the networks at that time, and Markham implied that "promoting a project with which the networks are out of sympathy" might be troublesome for the association. His suggestion was to start at the top by asking the networks what they thought about a code before contacting the stations. He then outlined a strategy for easing the industry into a mind-set receptive to a code. His plan involved Miller enlisting the support of the trade press and, at the upcoming NAB convention in Chicago, remarking publicly on "the deterioration of the television programming situation." Rather than push for "fixed commercial standards," Markham recommended that Miller's convention remarks advocate for a "living and growing document" and "a protestation of good faith on the part of industry where simple morality, good taste, and business ethics are concerned." The overt suggestion that Miller gently work the code into his speeches in order to prime the audience lends even greater weight to the rhetorical flourishes discussed in the previous section. Miller had already embraced the inflammatory rhetoric of the Red Scare. Now he was to build upon that with more pronounced specificity. The final piece of the strategy puzzle was the installation of network representatives on the Standards Committee to ensure network cooperation.

The same day that Markham sent Miller his memo, Theodore C. Streibert, president of independent station WOR in New York City, announced that his station would use the Motion Picture Production Code as a template for all of its programming, both filmed and live.[71] In his press release, Streibert argued that television and motion pictures' common formal characteristics made the Production Code "directly applicable" to the broadcast medium. Ultimately, WOR's use of the Production Code was a stopgap measure for Streibert, who stated that the industry needed to self-regulate and institute "its own code incorporating principles peculiar to television."

In a letter to Miller the very next day, Streibert encouraged the NAB president to think about a code, although he claimed that lack of "experience" doomed its creation.[72] He also reminded Miller that the TBA had entertained the notion of a code "for several years" but postponed drafting one owing to television's newness. Tastes evolve, Streibert wrote, so he understood the need to wait, but he also acknowledged that some content requiring "correction in the future" should "be stopped before [it went] too far." His example was the crime drama, a genre constantly under attack by viewers. He wrote, "We know from the movies that it is bad practice to show how a crime is actually committed. Even a simple thing like knocking a man out by hitting him over the head is never actually shown in the movies. So far, however, there seem to be no inhibitions on demonstrating methods of committing crime on television." The TBA had anticipated the applicability of the Motion Picture Production Code's guidelines on depictions of crime and had incorporated those guidelines into its 1945 code. In the absence of that TBA code, stations had only Hollywood's model to follow. The adoption of the Production Code, though not Streibert's first choice, was "less objectionable" than not having standards at all. He wanted the NAB to circulate guidelines or best practices to stations without committing to a full-fledged code. A code could come later, though no one ever stated precisely when, and for Streibert it would be more effective if it avoided the "pitfalls and problems" encountered by the NAB's radio codes.

Miller agreed with Streibert's recommendations and even added that progress in the area of standards would be sufficient to lure Walter J. Damm, practically an NAB institution, back into NAB membership.[73] Because Damm's station, WTMJ of Milwaukee, Wisconsin, had left the NAB, so had Damm. He had written to Miller on March 16 telling him that "a code of decency or standards of operation" would help convince the *Milwaukee Journal* stations to rejoin.[74] Although Miller was happy to go along with a proto-code, the NAB would have to contend with institutional obstacles. The combination of dues-cutting and the departure of major stations (like WTMJ) had put the association in an unstable financial situation. The NAB was diverting money to efforts like sales and advertising that were more outwardly advantageous to stations and away from areas with less tangible benefits. Miller needed Streibert to use his influence to persuade other station executives to support a standard-setting initiative. In less than two months, ABC and CBS would resign from the NAB, so the financial difficulties Miller sketched in his letter were not set to improve quickly.

Meanwhile, *Variety* reported on two codes that had been floundering for some time: the TBA's code and a code crafted in 1949 by the Independent Television Producers Association (ITPA).[75] The ITPA code—"a point-by-point system of regulations"—was delayed because of merger talks between ITPA, which was based in New York, and the Television Producers Association based in Hollywood.[76] In the midst of this news of stalled industry codes, Markham was able to report back to Miller on the results of the NAB's survey of its members' program policies.[77] Of the 109 queries mailed to its members, only 36 replies reached the NAB. Reporting the facts to Miller, Markham admitted that "the majority of stations are inarticulate on this particular subject, which, in turn, suggests that they have been slow in coming to grips with a somewhat 'touchy' matter." The results of the survey expose a situation uneven enough to make the NAB leadership nervous. Twelve stations had instituted their own codes, and only six had "reasonably comprehensive program policies." Markham relayed that the majority of stations had "established no specific TV standards" but were "applying the spirit if not the letter" of the 1948 radio code. Since most stations followed no code of their own making, they policed themselves on a case-by-case basis—a tactic supported by the TBA.

Within Markham's report were extracts from some station managers' comments that divulge the strategies and mind-sets employed at the local level:

> If a substantial number of complaints about . . . a program are received we
> knock it off the air. We . . . impose . . . censorship from the point of view
> that the average parent would take in regard to exposing his children to
> the program (WHEN, Syracuse); We have refused several beer and wine
> accounts. . . . We have refused certain network sustaining programs we con-
> sidered unsuitable. . . . We will definitely develop a policy list. . . . At the
> moment all we can say definitely is that we take the greatest care (KSL, Salt
> Lake City); [. . .] We are attempting to follow the same pattern of public ser-
> vice broadcasts that have always been termed acceptable and in good taste
> for AM (WOW, Omaha); We avoid mysteries before 8:30 pm. . . . All films
> are screened for harmful content, the same person also screening for five

other stations including WBTV, Charlotte, and WSB-TV, Atlanta. . . . We are being careful to control our programming at this stage. Reducing rules and regulations to writing would not be helpful at this time (WAFM-TV, Birmingham); [. . .] Our producers would not permit anything off color to go out. . . . We have no policy formulated; that seems to be a problem for the networks (WRBM-TV, Indianapolis); [. . .] Our policies are pretty much a projection of the policies adhered to previously in our radio operations . . . our agreements with the networks permit a rather wide degree of program rejection and acceptance . . . no beer, wine or liquor advertising will be accepted here (WOI-TV, Ames); We have tried to use the "common sense" approach. Whenever anything comes up in our programming which could possibly offend the audience, our policy has been to delete it. . . . Until some national code is worked out, we intend to continue with the "common sense" system (WTVJ, Miami); [. . .] Television is still new and we are changing our ideas so rapidly that we do not think the time is right to formulate fixed policies. . . . Our policies are being dictated pretty much on a day to day basis although in our minds there is a pattern which is being formed (WLW-T, Cincinnati; WLW-C, Columbus; WLW-D, Dayton).

The stations seemed to navigate the NAB-network-station triangle expertly if delicately, assuring the association that their eyes were squarely on their communities and, thus, on their licenses. Like parents watching over children, these stations exercised a great degree of control over their airtime by aspiring to the undefined goal of good taste. But they also notified the NAB that their locally originated programs were not of concern. Network programs were the problem. Markham learned that some of these stations had appealed to the networks to "monitor their shows" more "strictly." Most, however, were not overly critical of the networks, and Markham sensed this was a consequence of their dependence upon the networks for live programming.

Left to their own devices, the stations regulated the situation with varying degrees of severity. The six stations that did implement policies targeted similar problem areas. Common restrictions targeted commercial time, alcohol advertisements, graphic crime, vulgarity, and programs that aired when children were awake. Some deviations from these standards point to the region-specific cultural concerns of local stations. WTVR in Richmond, Virginia, prohibited "programs in which the races are mixed." WBEN-TV in Buffalo, New York, forbade the "degrading or ridiculing of organizations" that devoted themselves to "improvement in morals, spiritual life, and conduct." Other policies indicated more commercially minded considerations of programming flow. WFMY-TV in Greensboro, North Carolina, encouraged "'mood' programming" and restricted the quantity of any "one type of entertainment per night or per week." Still other policies ensured that stations would skirt illegal activities. For example, WMAR in Baltimore aired "horse racing results under circumstances which [made] them valueless to bookies and gamblers." Only one of the six stations, WTGN in Minneapolis, used the Motion Picture Production Code as its guide.

Was the Production Code a reasonable template for television stations? Some were looking to Joseph Breen, who administered the Production Code, to be the authority on its compatibility with television. Breen would broach that topic on April 21 at an American Television Society (ATS) event.[78] Reacting to this news, Markham sent a memo to Hardy, asking him to attend Breen's talk.[79] Unable to attend himself, Markham worried that the TBA and the ATS were appearing "to get the jump on NAB when it comes to formulating standards." The TBA would have posed a greater threat to the NAB, as ATS was a smaller organization operating less as a trade association and more as "an intelligence center and clearing house for information pertaining to television," as well as a "forum for the exchange of ideas and discussion" of television-related issues.[80] Given the NAB's on-again-off-again merger negotiations with the TBA, the larger association did not want to be seen as lagging behind the TBA on such an important issue as standards.

After Breen's talk, NBC censor Stockton Helffrich had much to say about the event in a memo to Pat Weaver.[81] In attendance at the "closed session" were representatives from ABC, WOR, and a number of "other interests not identified." Helffrich suspected that they were somehow affiliated with the American Television Society and that much of the conversation was rigged according to the wishes of the ATS. The ATS, in Helffrich's assessment, wanted a "television parallel" of the Production Code, wanted "the networks to fall in line," and, finally, wanted "a finger in the pie." NBC's censor dismissed Breen as "pretty inept" and saw through the Production Code administrator's rhetoric. He regarded as hypocritical Breen's statement that the Production Code did "not reduce to a common denominator artistic effort," and he further criticized the Production Code Administration's empty, public relations–fueled gestures and lax applications of its own "detailed [. . .] rulings."[82] In sum, Breen was wholly unsatisfactory to Helffrich, who wrote that Breen purported to have "all of the answers" when in reality he did not have "any more answers on all of these problems than do broadcasters." Helffrich pointed to the singular rupture between Breen's job and his own at NBC: "Those problems in his field are confined to motion picture dissemination and in our field to radio and television audiences." The matter of theatrical distribution versus live production, distribution, and reception was a key distinction for a television censor, and that distinction would support the greater applicability of a radio code for television than a motion picture code. Pat Weaver stated as much in his reply to Helffrich, questioning the expertise of people like Breen: "I do not know why anyone thinks the movie people understand the rules of good taste, etc. in programming for the all-family audience at home. Only radio has met and knows this problem. And radio-trained men should solve it. But we must solve it, not let it get away from us."[83]

Once again the home became the focal point for those who sought the correct path for regulation. However, while home viewing was on everyone's mind, its meaning and implications lacked stability. Helffrich seemed to be concerned with the maintenance of artistic integrity within the constraints of the home, but Weaver set his sights solely on "good taste." In Markham's query to stations, local broadcasters sometimes catered to home-based viewers in culturally specific ways, thus conceiving of the home as an enclave for conserving community norms. For the TBA in 1945, home demanded decency and good taste, but it could also be a space for educational uplift. Each of these preoccupations contributed to a sense of urgency around what would be the principal site for television viewership, but to what extent the resulting code would address that urgency with medium-specific provisions remained to be seen.

Whether or not the Motion Picture Production Code was the correct model for TV, Markham ascertained that the majority of stations wanted standards.[84] Creating those standards would be a difficult task, to say the least. "Nevertheless," Markham wrote to Miller, "if we do not begin to protest our good faith and awareness of our social and moral responsibilities, some opportunist may suddenly make life even more difficult for us by forcing down our throats some legislated form of censorship and some standards which are altogether unrealistic." Markham failed to elaborate on what unrealistic standards might look like, but evidence suggests that these may have included anything involving fewer advertisements and more educational programs. The solution as Markham saw it was to announce the formation of a television standards committee and commence drafting a "modest" code that would be revised each year. Annual revision would allow the NAB to stay ahead of government action.

TBA leadership did not want the problem to get away from them either. Perhaps as a response to the activity of the NAB and the ATS, the struggling trade association established a new committee to study programming early in May 1950.[85] That the committee's chair was Lawrence W. Lowman, vice president at CBS, was probably an important sign, since CBS would resign from the NAB later that month. The committee's task was to revisit the TBA's 1948 "Statement of Principles" with the benefit of greater "experience." Meanwhile, by mid-May, Miller had met with the four networks and reported to the NAB board of directors that the network heads felt it was too early to draft a code.[86] Charles Denny from NBC, Joe Ream from CBS, and Robert Kintner from ABC all felt that the networks "should have a much larger voice in the determination of standards [. . .] than would be true if NAB undertook the job" at that moment. Miller questioned the wisdom of heeding the networks' insistence, given CBS's and ABC's "defection," as he put it.

As discussed in chapter 2, the fall of 1950 saw a scramble to incorporate television into the NAB, and by January 1951 the NAB was confident that most television stations would approve of NAB-TV, to the detriment of the smaller TBA. Indeed, the stations that attended the NAB's January 19 meeting in Chicago voted in favor of the new venture, which would officially launch on April 1, 1951. The programming problem did not simply disappear while the NAB absorbed the TBA's television operations. The trade press warned of impending government intervention. An article in *Television Digest* encouraged readers to "start boning up [. . .] on the radio and motion picture industry codes—and better get your trade groups working on something such for TV. Otherwise, Uncle Sam may really slap you down—and with plenty of popular backing."[87] Speculating about the FCC's upcoming programming conference, the article warned that the commission "intends to go further than mere questions of 'program balance'" and would tackle "such sure-fire headline-provoking aspects as off-color jokes, plunging necklines, [and] crime dramas during children's viewing hours. They're even talking about arriving at some sort of definition of 'taste.'" Not all FCC commissioners agreed that such oversight was within their purview. The article cited Commissioner Robert Jones's opinion that the FCC should remain outside of programming matters and focus only on granting licenses to the "right" applicants, a process that actually had a great deal to do with programming matters, as discussed in chapter 1. The highly anticipated FCC programming conference never materialized, but the possibility of such a conference performed two functions: it hung over the newly renamed NARTB as a threat and simultaneously reminded ambitious television reformers that the FCC was aware of the situation and was capable of handling it without congressional action.

Movement toward a code accelerated as 1951 proceeded. NBC's Continuity Acceptance Department formulated a new policy manual tailored to television early in the year.[88] According to Robert Pondillo, the *NBC Radio and Television Broadcast Standards* "was considered the freshest, 'most comprehensive' collection of self-regulatory TV standards available," and it was instrumental to the drafting of the Television Code.[89] It did little, however, to dissuade elected officials like Representative Thomas Lane (D-MA) or

Senator William Benton (D-CT) from introducing resolutions and legislation to study television and monitor it. Benton's S. Res. 127, which called for a study of television programming and alternatives to the commercial system, coincided with the NAB's new life as NARTB in April 1951. As the NARTB contended with what it perceived to be attacks from lawmakers, it underwent a leadership change and set about drafting its code. Harold Fellows replaced Justin Miller as president in June 1951, and Miller stayed on as chairman of the board.[90] Also in June, the NARTB Television Program Standards Committee met for the first time and resolved to adopt "'standards of self-regulation' designed to improve the character of television programming and insure its observance of good taste."[91] Both the NARTB and television reformers had gained momentum.

Part of the NARTB Standards Committee's charge was to

> concern itself not only with the day to day program problems of television but with the broader aspects of its effectiveness as a mass medium, its impact upon public morals and morale, its effects upon the welfare of the family and individual members thereof, with particular reference to children, its contribution to the cultural progress of the nation and its influence for the good upon the behavior patterns of American society and the society of nations.[92]

At least discursively, the committee saw content not as images isolated on a screen, but as a participant in the development and maintenance of families and citizens. Television was not just an appliance; it was a potential authority figure in the national family, but that figure required an education to determine how best to execute its authority. Anna McCarthy explores the genesis of this thinking and argues that the television industry was keen to consider the role of television in propagating additional forms of governance and self-governance. TV could be "a mechanism of *security*" that would transform entertainment spectatorship into "a form of conduct that could, under the right conditions, allow the population to reflect upon the social realm and form opinions, regulating itself automatically without sacrificing freedom of choice."[93] Dovetailing magnificently with the commercial broadcasting industry's history of self-regulation, this focus on self-governance, as McCarthy notes, reinforced the "understanding of corporate leaders as moral guardians."[94]

Such an understanding saturated the code discussions. The burden of moral leadership that the committee placed upon itself—or at least that it declared it placed upon itself—resulted in additional committees and a drafting period completed in relatively short order. Only three months passed between the formation of the drafting committee and the adoption of the Television Code by the NARTB membership.[95]

Of course, NBC's standards and the NARTB's Television Code did not completely solve the problems they purported to solve. In 1951, an angry memo from Joseph McConnell at NBC to everyone involved in the network's television production illuminated the tensions simmering over content, even though NBC had standards in place well before the Television Code was enacted.[96] "I am sick and tired of receiving justified criticisms of NBC television programs where bad taste is concerned," McConnell wrote. Placing final responsibility in the hands of the director, McConnell faulted everyone who came into contact with a program, from its development to its airing, for "dubious" material. If the program would not be acceptable in the employees' homes, he reasoned, then it should not go on the air. A summary threat concluded the memo: "Any borderline material not questioned from here on in, and subsequently the target for public censure, will be the cause of considerably more than censure from your company's Management for the personnel responsible." NBC content needed to be a welcome guest in the home, but executives like McConnell considered it an unruly creature that required multiple minders and rigorous discipline. Above- and below-the-line talent could lose their jobs over the success or failure—as determined by ratings—of that creature, but also over the creature's taste level, an imprecise yardstick fashioned at that moment by censors and multiple committees.

The NARTB Pats Itself on the Back

Approved in October 1951, the Television Code helped the NARTB get out from under threats of government intervention, the breadth of which will be discussed in chapter 6. But in a speech to the ATS in late November, Thad Brown, NARTB's director of television, proposed an additional reason for the Code. For Brown, "pioneer telecasters" wanted to demonstrate their "voluntary sense of responsibility [. . .] and guarantee [. . .] a wholesome stature for the commercial television broadcast industry." NARTB president Harold Fellows also resisted the typical reading of the Code's adoption in a speech to the Television Association of Philadelphia: "I believe television broadcasters wrote and adopted a code because inherently [. . .] they are decent, self-respecting, God-fearing citizens. Television is going to be as good as you are [. . .]; in your absence it's going to be as bad as the Government can make it—and that can be [. . .] pretty bad."[97] After celebrating the character of broadcasters, he returned to the rhetoric that had served Miller so well for so long.

"We adopted this Code before someone decided to adopt one for us," he said in a speech in March 1952—the month the Code formally went into effect.[98] Rewriting the reason for the Code was essential to the NARTB's propaganda efforts. They were not forced to do it; they did it voluntarily for the good of the nation and for the good of TV.

As Mark MacCarthy notes, "Industry codes sometimes have the public re-lations function of creating a favorable impression of a responsible industry policing itself."[99] The Code as a public relations document, one that superfi-cially showed the public the industry cared, lent a new dimension to Fellows's speeches as he gradually began to incorporate the public into his portrait of the association and its battles with the government. A speech in May 1952 en-titled "Liberty—Let's Keep It" pit the government against the nation's citizens and cast the NARTB as the custodian of democracy. Fellows remarked, "Judg-ment, program tastes, preferences, and economic decisions affecting broad-casting, under our American system, belong to the public—ALL of the people, and not to the government."[100] He went on to argue that by choosing which programs to watch, the people were exercising their freedoms. In a fiery speech in late 1952, Fellows extended this argument, claiming that all attempts to con-trol television were offenses against the people.[101] The Television Code, which controlled many aspects of television, was not implicated in those offenses.

For Fellows, the Code was also proof that the industry was "grown up now," but he still referred to the NAB-government relationship as one of marked abuse.[102] "Beyond the garden lies the woodshed," he stated. "That's where you get whipped! Over and over again—for thirty years—we've been constantly summoned to the woodshed by all kinds of people, for all kinds of reasons. By now we know too well where the woodshed is and who's in there. From here on when we're summoned there without cause . . . let's ask the gentle-man with the birch branch in his hand to answer a question or two before we bend over to take it."

Fellows's "Liberty" speech came just a few weeks after Miller commented in a letter about the maturation of broadcasters. He wrote that broadcasters' guilt over poor programming initially led them to seek out "government disci-pline" despite the possibility of a "gradual government abridgment" of free-doms.[103] The Code signified a change; "broadcasters are beginning to 'grow up,'" as the press did, according to Miller. The purpose of the letter, how-ever, was to raise awareness of what Miller hypothesized could materialize: a constitutional amendment removing First Amendment protections from the broadcasting industry. By invoking this possibility, Miller leapt over Blue Book–style regulations and imagined an extreme outcome perfectly in keeping with the angry rhetoric of his speeches. The "strength" of television, for Miller, was its pervasiveness and direct connection to the "family circle." However, that connection became television's "weakness" because it caused people to forget that broadcasting was protected speech. Only the industry's "ability for self-government" had fended off the extreme possibilities. For Miller, home was ultimately a detriment to the freedom of the TV industry. Home represented millions of consumers, but these millions of consumers had something to say about the images and sounds that entered their homes.

Conclusion

While the NARTB argued against government censorship legitimately—the Communications Act of 1934 expressly forbade it—the association's leadership insisted on projecting an image of abject victimization at the hands of an oppressive regime. Did they really believe it? As I will discuss in chapter 5, internal NARTB documents reveal a much more complicated relationship with the government than the speeches would have us believe. Yet, while there was a willingness to give a little to get a little in the mid- and late 1940s, by 1950 the experiment of television had inflamed anxieties about content. Some began to seek out new paths to government involvement in that experiment. Public and private actors with varying degrees of power scrambled to fix problems before they got out of hand and before television became immovable. Potential reformers like Senator William Benton, a former advertising executive of all things, did not want to lose time while the industry dug in its heels. The industry, for its part, resorted to scare tactics and mudslinging to protect its commercial way of life.

As chapters 2 and 3 have intimated, the march to a code for television might have proceeded down a different route had the TBA (and not the NAB) housed television permanently. The TBA, also subscribing to the ideologically problematic ideals of wholesomeness and good taste, had flirted with codes since 1945, but its willingness to endorse station autonomy delayed any formal adoption of a code. Where the story veers into even more problematic territory is in some factions' embrace of the Motion Picture Production Code. Stockton Helffrich believed that the conditions of television's reception bonded so intimately to its content—the family circle and the acceptability of its live entertainment being profoundly intertwined—that TV's adoption of a movie code was unacceptable. Television's most visible censor did not want to see the medium's talent reduced to a common denominator, as motion pictures' most visible censor, Joseph Breen, had done. Helffrich also did not want a television code of standards to be an empty gesture. An examination of NBC's

Continuity Acceptance Department reports, like Pondillo's, reveals Helffrich's nuanced approach to content and helps explain his intensely negative reaction to Breen's placement within television circles. In the end, as mentioned in this book's introduction, the Television Code did pull from the Production Code, the NAB's 1948 "Standards of Practice," and NBC's standards. Welding together key parts of these documents signified that the easiest way to regulate television content was to disregard the TBA's claims of technological and industrial singularity. The old habits formed by decades of radio codes quashed a truly new approach to self-regulation.

The different approaches to content detailed in this chapter—local broadcasters' protectiveness of their airtime, the TBA's unwillingness to commit to a code, the ATS's enthusiasm for a transmedia approach to control, NBC's almost proprietary claim on radio- and television-specific standards, and the NARTB's obsession with shielding the commercial system from government-based reformers—amount to something greater than ego-driven clashes. Their objectives were different because their stakes were different. Licensed versus unlicensed, national versus local, sight versus sound, filmed versus live, theater versus home, regulated versus unregulated—these binaries weighed on television content *within* the capitalist organization of media. The industry was not monolithic by any means, though its pursuits were certainly predictable within the logic of its commercial enterprise. Distanced from the mechanics of the battles over content, yet engaging with the outcomes directly, was the TV audience. In the next chapter, we will see how motivated audience members drew a line from their television programs directly to seats of private and governmental power and in the process embodied the public interest, convenience, and necessity.

CHAPTER 4

The Television Audience Speaks Out

Viewer Complaints and the Demand
for Government Intervention

In 1952, a young girl named Mary Boston from Colorado wrote to President Harry Truman asking "that the radio programs for children, such as *Big John and Sparky* and *Mark Trail* be made longer."[1] The letter, forwarded to the FCC, was treated to a reply by Secretary T. J. Slowie, who broke down the complicated system of US broadcasting for Mary. Slowie wrote:

Mary, while you may be too young at the present time to understand the full meaning of what I am going to say, I will try to explain briefly, broadcasting as it is done in our country. In America, broadcasting is under private enterprise. This means that the man who operates a radio station selects the programs that are to be broadcast. He receives a license from the United States Government which gives him permission to broadcast only on a certain frequency with a definite amount of power. The hours during which he is permitted to broadcast may be set by the government or, if no interference will be caused to other stations, he may broadcast as many hours each day as he wishes. [. . .] The government requires that he operate the station in the public interest. That is, he must broadcast programs that will serve the different interests of the people living in his city. The government also expects a station to broadcast programs which will assist in meeting the educational and civic needs of his city. To do this, a station should broadcast a certain number of programs in which children are particularly interested. He should also broadcast programs in which the other citizens are interested. To prevent any possibility of the Commission taking over powers contrary to those intended, the Congress stated that this Commission should have no power of censorship over radio programs. The Commission reviews a station's over-all operation at certain times to determine that it has been operating in the public interest. However, it cannot tell a station what programs to broadcast. In some other countries, Mary, the people can hear only the programs that the government wants them to hear. If you wish, you may write letters similar to the one sent to the President, to the broadcast stations in Denver, Colorado. Enough letters from children might result in more programs of the type you are interested in hearing broadcast. However, you must always remember that a radio station broadcasts programs for everyone and not just for children.

Children were (and continue to be) the symbol of everything innocent that required protection from the flow of broadcast signals. Their special status even warranted customized guidelines in both the radio and television codes. But as a viewer staring at a screen and wanting to change something about it to suit her preferences, Mary Boston is actually the ideal stand-in for the television audience, regardless of her age. Her ignorance of the regulatory system was shared by her much-older compatriots, yet, unlike other disgruntled viewers, her young age afforded her a reasonable, if flawed, explanation of how television came to her. She was even treated to a definition of the public interest. Adult letter writers were not extended the same courtesy. They were treated as though they should have already known how television worked, as though the complex system of legislation, administrative agencies, self-regulation, production and scheduling control, advertising agency involvement, and network-affiliate relations required no explanation once viewers reached a certain age.

Many stakeholders traveled the road leading to the Television Code, and none was more powerless than the television audience. Having no investment in the television industry, viewers were equipped with two resources: their purchasing power and their citizenship. Viewers incensed at any number of television-related issues complained to networks, stations, and sponsors as consumers. They complained to the government as owners of the airwaves.

Although the US system of broadcasting was not conceived as a government-run enterprise, its early architects hesitated to relinquish ownership of the electromagnetic spectrum to private companies. The solution would be a compromise, or what has become known as the "trusteeship model" of broadcasting. The public owned the airwaves, and the government, with the public interest in mind, selected which broadcasters could operate on a limited number of frequencies. Awarded a license, for which they paid nothing, private broadcasters were free to convert their investment in a station into profit from the sale of advertising, provided they operated in the public interest. These broadcasters were simply holding on to those frequencies as long as the government said they could. As trustees of the airwaves, they needed to follow rules and serve their audiences. When they failed to do so, their audiences took up pen and paper.

Most people who provide feedback are very motivated to do so, and as a result their comments are either overwhelmingly positive or overwhelmingly negative. Because the FCC solicits complaints and not praise, the letters in the commission's papers at the National Archives range from sensibly critical to hyperbolically outraged. The analysis in this chapter draws from 1,084 letters written between 1948 and 1952. The letters were read and classified according to their dominant themes (complaints) and the gender of the writer (when discernible). Most of the letters studied here were written by viewers to the FCC. Some letters are replies from the FCC to those viewers. In a few cases, viewers wrote to their congressmen, stations, networks, and even, like Mary Boston, to the president of the United States. In these instances, either the writers sent courtesy copies to the FCC or the original recipients forwarded the letters to the FCC. Evidence in the FCC papers indicates that, with a few exceptions, the commission replied to each letter in a standardized fashion. This set of standard replies, known as a "performance program," is necessary when an organization is routinely beset upon by "demands for change."[2] The organization needs to "anticipate the disturbances before they disrupt normal operations," so it implements a process triggered, for example, by a common complaint.[3] Most FCC replies began by briefly restating the reason the viewer sent the complaint letter. The reply went on to explain that Section 326 of the Communications Act of 1934 forbade the FCC from guiding or censoring content. This strategy—what Joseph Turow calls "rule invocation"—is designed to align the law governing the FCC with the viewer's own commitment to the First Amendment, thus neutralizing the foundation of the complaint.[4] Neutralization was the goal, but most likely it only frustrated viewers who sought action.

The exercise of reading complaint letters outside of their original context is a unique one. On a superficial level, many of the letters are humorous because they reflect sensibilities that seem prudish by contemporary standards. The writers often sound naive and sometimes childish. Some get aggressive; others are gentle but shocked at TV content. Thankfully, very few (but still too many) funnel their anger at TV through bigotry. A lot of them worry about their children. Ultimately, in examining these letters, we are reading something not meant for our eyes, but we are recognizing something that was important to the letter writers. As Charlene Simmons notes in her study of fans' perceptions of interactivity in early radio, one valuable aspect of archival letters is their representation of a moment uncorrupted by memory or analysis. She writes, "The letters contain the thoughts and perceptions of a variety of radio listeners without the outside influences of historical recall, as would be present in oral histories or texts written after the passage of a substantial period of time."[5] The concentrated and sometimes unfiltered subjectivities articulated in these letters may not represent the feelings of all audience members, but they can point to patterns and norms of particular demographics.

In a material sense, the letters represent a segment of the population that was wealthy enough to own a television set and had the home-based employment or the leisure time to engage with it. References to money spent on expensive television sets and disruptions of family life indicate that most of these letter-writers were not complaining about programs viewed in a public space, such as a tavern. The letter-writers knew which institutions to contact, although their knowledge of these institutions' powers varied. Most of the writers were women, and most of their concerns centered on the welfare of children and the need to shield them from violence and sex. By far the greatest number of letters targeted the indecency and crime prevalent in television programming. But other subjects, such as sports blackouts, advertising, and religion, made frequent appearances. Apart from identifying and analyzing the mainstream and marginal preoccupations of these letter-writers, this chapter connects viewers to the themes in the rest of the book by exploring five areas: (1) viewers' interaction with the processes and circumstances surrounding the industry's and regulators' insertion of television into daily life; (2) the concerns of viewers that relate directly to the eventual terms of the Television Code; (3) viewers' common pleas for government censorship; (4) the responses of some in the television complex (the regulators and the industry) to viewers; and (5) the viewers' response to the Television Code. By considering the perspective of viewers who chafed at trends incompatible with their understanding of what television should be, we can broaden our understanding of the relations behind the Code.

Viewers versus the TV Complex

In October 1952, Edith Obstfeld of New York City wrote the following to the FCC: "Having accidentally tuned on to *Strike it Rich* this morning, I rushed to the World Almanac to find out what the obligations and powers of the FCC are. The Almanac doesn't say. I'm angry enough to want to know!"[6] Viewers who wrote to the FCC commonly understood that the commission was the one to call on, but just as commonly were not quite sure what the FCC did. To be sure, confusion about the FCC's powers is commonplace in general, as the next chapter will explain. But whereas scholars disagree over the scope of the commission's authority, laypeople like Edith Obstfeld simply wanted to determine how the FCC could help them navigate the new television landscape.

Ms. Obstfeld was operating in a way not encouraged by the architects and practitioners of commercial broadcasting. Audiences are commodities in this system, but they are also constructed as consumers. As Turow reminds us, advertiser-dominated media have created a landscape in which it is not beneficial for media companies to construct or even imagine audiences as "producers of a civil society."[7] Rather, audiences should "think of themselves primarily as leisure-oriented consumers of media content and sponsored products."[8] By contacting the government, Ms. Obstfeld and other viewers disrupted the identities and behaviors that networks and advertisers had manufactured for them. The first step to reform may have been to circumvent the obvious industry players, but the next step was to ascertain how (or if) the government could intervene.

Former NAB board chairman John F. Dille Jr. opined on the matter of public input, no doubt informed by the NAB's history with it. He wrote:

> Broadcasting is a peculiar, unique kind of business affected, in the regulatory sense, more by "climate"—political, moral, and social—than by legalities. Broadcasting is never merely a dialogue between the regulators and the regulated. A vitally important third party praises, carps, kibitzes, criticizes, suggests, and generally makes its opinion known in a variety of ways—the entire American public.[9]

Dille cautioned that these comments could not be generalized; instead, they were "highly individualized and personal and at the same time extremely diverse."[10] Nevertheless, scholars agree that the public is a potential thorn in the side of regulators. One "downside" for government regulators, according to Clarke, is the probability that "public expectations" will be higher than if there were no regulations in place at all.[11] Expectations would turn into dependence, as people would eventually "rely on state regulation to achieve their complete security and so fail to take prudent measures themselves."[12] In some ways, this scenario in a broadcast regulation context is ill-fitting; audiences' sheer lack of institutional control drastically minimizes their ability to effect change. In fact, if audiences are the product in the commercial broadcast scheme, packaged by measurement companies according to the desires of advertisers, one could hardly anticipate grassroots reform. But Clarke's scenario can help explain the vigor with which audiences, feeling ignored by the industry, looked to the government regulator to act on their complaints and reprimand those who would shirk their responsibilities to the public.

Misunderstanding the FCC

Viewers' belief that the FCC could censor, which will be explored later, was one of the most significant points of confusion in the archived complaint letters. A less constitutionally problematic but more industrially interesting misunderstanding centered on the government's basic connection to programming. Two letters to President Truman, which the White House forwarded to the FCC, appealed to the president to lengthen Liberace's half-hour television program. "After all, you're the president," one woman wrote.[13] Members of the *Tom Corbett–Space Cadet* fan club and their parents wrote to the FCC to ask that the canceled program be put back on the schedule.[14] One of the adults even piggybacked on the *Space Cadet* cancellation and asked for the reinstatement of *One Man's Family* and *Ted Mack's Amateur Hour*.[15] The removal of *Roller Derby* from the air also provoked a number of letters.[16] In reply to one married couple, FCC secretary Slowie wrote, "The Commission has no authority to require broadcasters to carry or not to carry a particular program."[17]

In addition to believing that the FCC could modify programs or even revive a canceled show, some viewers thought that, like the networks' continuity acceptance departments, the FCC monitored programs in advance of their airing. Upset by a drunken comedy routine, one viewer asked the commission, "Did your monitor not think that the program [. . .] merited at least a warning to the offending station?"[18] On another occasion, a US senator wrote to FCC chairman Wayne Coy on behalf of a constituent who wanted "to receive a commission from the FCC to act as a monitor to check by use of a stop watch this time limit" on advertisements.[19] Far from being outliers, these letters exemplify the tone of most of the complaints. Even though most letter-writers did not offer their services to the government, they behaved, in effect, as a neighborhood watch program.

The Audience and Televisual Processes

As viewers settled into the routines of television, they also had to contend with process—the way television would be produced and distributed. Understandably, viewers' expectations of liveness spilled over from radio into television. Live programming has been valorized as the essence of television and the primary element that distinguishes broadcasting from film. In his work on midcentury TV, Boddy discusses 1950s critics' preference for the immediacy of live television over film's "feel of the past."[20] But as Jane Feuer and others have argued, the construction of liveness is ideological and promotes "flow and unity" over the reality of fragmentation.[21] Production practices interfere with the flow of any event, spoiling the impression that the event is unfolding before viewers' eyes. Nevertheless, industrial practices in early television were instrumental in branding live TV as better TV. Just as with the radio networks, local stations had to affiliate with television networks if they wanted access to live network programs.[22] For the networks, live TV production, based in New York City, was key to building their affiliations and combating the threat of filmed programming coming from Los Angeles. Stations did not need to affiliate with anyone to procure programs from telefilm producers. Liveness was therefore an early cornerstone of the four television networks' dominance over national television distribution.

Liveness also endured for regulatory reasons. Sewell notes that David Sarnoff connected the communal experience enabled by live broadcasting to "public service in the national interest," but that was one executive's opinion.[23] More impactful was the Federal Radio Commission's 1928 pronouncement that liveness ("the broadcast of events or entertainment unavailable to the public in any other form") served the public more than a recording would.[24] The power of the FRC's statement endured into the television era. Filmed programs had no accompanying scarcity, so they were comparatively less important.

For audiences, however, the importance of liveness to public service or to the maintenance of the network model was beside the point. They had grown accustomed to live programming on radio, and they expected the same from TV. "The owners of T.V. sets are taking an awful beating on what's live," one viewer wrote, further suggesting that live programs be marked "LTV" to distinguish the real from the "fiction."[25] Another writer berated "'dead' kinescope television," particularly with regard to sports, and he suggested using filmed television images only to replay game highlights (with "touchdowns in slow motion").[26] Not all viewers loved liveness, or at least the antics of live studio audiences. In a letter to the networks with a courtesy copy sent to the FCC, one frustrated viewer wrote, "It is neither public interest nor necessity to be forced to wonder what a studio audience is laughing about, nor to watch Aunt Minnie and Uncle Ben waving to the family."[27] Suffering under limited television service during the license freeze, another viewer complained that his lone TV station still did not support live television.[28]

Kinescope broadcasts—live programs recorded on film for distribution to non-interconnected stations—irritated a few viewers, one of whom believed that they were not worth the money he paid for his TV set.[29] Reruns, too, irked the price-conscious viewer who felt pressured to keep up with the bigger and costlier sets. A greater investment in technology was not rewarded with fifty-two weeks of new live programs, so one viewer, calling reruns a "racket," begged the FCC "to do something about actually punishing any organization that forces us to see its shows repeated."[30] In his study of the industrial and cultural significance of reruns, Derek Kompare explains that even though the practice of airing recordings on local radio had overlapped with the "ideal of live programming" on the networks from the 1920s through the 1940s, reruns were still somewhat suspect.[31] But if the fight over recorded programs was at least dying down on radio, television presented an opportunity to revisit the issue. "The earlier aesthetic and economic debates about the validity of live vs. recorded programs were repeated nearly verbatim" on the occasion of television's mainstream launch, according to Kompare.[32] Complaint letters give us a view of those debates from viewers' living rooms.

Along with kinescopes and reruns, movies on television posed a problem for viewers. Angry that films were taking up space better occupied by television programs, one viewer called the broadcasting of films "unfair" to TV owners and insisted that "it should be stopped."[33] This viewer wanted television to be purely television, but others were irked by the quality and frequency of the films. The market for movies on television was severely limited because of what the US government believed was conspiratorial behavior. In 1952, the US Department of Justice filed suit against multiple producers and distributors, including Twentieth Century–Fox, Warner Bros., and RKO, alleging that these studios had withheld their library of 16mm films from television to stave off any threat from the new medium.[34] Jennifer Porst explains that although 35mm was the standard gauge for theatrical releases, most studios transferred their films to 16mm for exhibition by the military and various aid organizations.[35] As military demand for these films dried up after the war, television represented a new market. Although the transfer from 35mm to 16mm degraded the film's image, the "lower-resolution 16mm film was perfectly adequate for smaller, lower-resolution television screens," according to Porst.[36] Regardless of the compatibility of 16mm with television screens, the studios argued that selling films to television during the license freeze, with so few stations on the air, would have come at a cost. But as Porst points out, internal correspondence in the late 1940s clarified that money was not entirely the issue; some executives simply did not want their films on television.[37] However, once more stations went on the air, the exchange value of the studios' libraries would increase substantially. In the meantime, the studios also worried that exhibiting films on television would decrease their value if re-released or remade.[38] Union and guild demands for residuals from television exhibition also helped to shut down the conversation. As a result of a myriad of circumstances, then, until the mid-1950s television had a limited supply of mostly "independent and foreign films" to broadcast.[39]

That scarcity turned up in a complaint from a Chicago viewer confronted with "the same movies on different stations time and time again."[40] Another viewer asked, "Is there a gross fraud being perpetrated against television set owners in Baltimore?"[41] He claimed that he was promised "direct entertainment" when he bought his television set, but in reality his viewing options were "principally poor specimens of films." Decrying the repetition of older films on TV, one writer called for the "elimination of these bleached and washed out, decrepit, broken down films," and another sarcastically suggested that movies older than five years be classified as "historical film[s]."[42] A Philadelphia man complained that he had seen the same older westerns "five to ten times a year."[43] Prefiguring pay television movie channels like HBO, he proposed the creation of a "Public Movie Department" that would "charge the TV owners ten dollars a year" and "purchase the rights of a movie outright," enabling it to broadcast newer, commercial-free movies. As Sewell argues, liveness and quality intersected discursively in complex and layered ways. For the FCC's letter-writers, the issue was comparatively simple: anything other than live or new was not worth the price of a set.

Production processes were another point of consternation for some viewers who were understandably unfamiliar with the demands of creating television content. As a result, one viewer believed that a cooking show was "practicing fraud" because the cook "display[ed] as her own creations food that she never cooked at all during the program."[44] The supposed spontaneity of man-on-the-street interviews on *We the People* likewise struck a viewer as fraudulent and insulting: "I have lived in N.Y.C. for over forty years and I know Times Square. Try and find five people so well versed in one group and so ready with the much needed gift of gab."[45] Letters like these are evidence of viewers well acquainted with radio and cinema but still feeling their way through the realities of television production.

Getting and Living with Television

The set, receiving a transmission from a station and operating loudly in the home, demanded that viewers adjust their habits and lifestyles. Each component of television most relevant to the home audience—the technology, the broadcaster, the sensory experience, and the domestic sphere—forced viewers to acquaint themselves with a new mode of daily life if they chose to complete the purchase in the first place. We have already seen that for some viewers the expense of the television receiver felt unequal to the quality of programming received. Anticipating that color broadcasting would debut "at [the] expense of television set owners," an anonymous viewer suggested that the networks first "show better programs throughout the day."[46] Part of the problem facing both networks and viewers was the dearth of stations for many communities around the country. The license freeze, which began in 1948 and did not end until 1952, halted the approval of new licenses, so many people who had television service at all received only one station. By the end of 1950, for example, 107 stations in 63 markets serviced approximately 9.8 million television sets, but 42 cities had only one television station.[47]

In 1950, with WDTV the only station broadcasting in Pittsburgh, one resident was motivated to ask the FCC to "agree we poor people are certainly entitled to a few more channels right now and without delay."[48] Fed up with subpar programming, a writer in the Pacific Northwest asked, "Why must the almost half million set owners suffer with KING-TV?"[49] Another viewer in Pittsburgh lamented her options "when something not to [their] taste" aired: "We can watch it, whether we like it or not, or we can turn it off."[50] Both she and a fellow viewer from Pittsburgh were complaining specifically about Arthur Godfrey, a popular variety show host and pitchman known for his improvisational style.[51] For one father, having only one station meant that his children accidentally caught sight of *You Bet Your Life*, or as he referred to it, "the putrid DeSoto Plymouth–Groucho Marx Show televised by KSD-TV."[52] Indecency and vulgarity were not the only concerns of people served by just one station. One viewer complained that after "going without" to purchase a television set, the only viewing options on the area's lone TV station were baseball games.[53]

Having surmounted the challenges of getting the television into the house and accessing a transmission, viewers next had to wrestle with assimilating the device into their sensory realms. Noise turned out to be bothersome for some of these writers. One woman, "kept awake till midnight every night by booming volume in the neighborhood," wanted the FCC to require all stations to "broadcast a spot announcement at eleven o'clock in the evening for set owners to consider their neighbors and reduce volume."[54] T. J. Slowie responded that the FCC had "no authority" to do so.[55] A letter detailing the anguish caused by excessive noise described how television was a "nerve wrecker."[56]

> On all speaking (including the stories, etc.) the back-ground music blares much louder [than] the speaking voice. Therefore where a program is adjusted for the speaking voice; during the pauses, the music blasts out from the back-ground, doing great injury to the nerves and effect of the whole being. I have heard of threats by neighbors, when other tenants play at a normal speak range, hit the ceiling when the back-ground of nerve-wracking music is filled in. I have seen radios and T.V. actually knocked over by the

haste to tone down the music. I have seen fathers actually brutally boll-over their children, who listen to the various stories, when the music is blasted in between the speaking.

A similar letter asked the FCC to fix the volume problem (even if it meant having a "constant volume law for everyone's delight").[57] Citing the commission's sound requirements for stations, the reply letter assured the recipient that, while "careless" stations are penalized for flouting the rules, our ears can deceive us. "The character and style of music used [. . .] can give a sense of loudness that actually does not exist." Truly, for the FCC to take into account mood music and aural flourishes—stylistic decisions, in other words—demonstrates the worlds the commission had to straddle as regulators of a creative industry. Beyond the sonic embellishments noted earlier, the FCC also had to contend with complaints about the elevated volume of advertisements, which, as two different writers asserted, "violate[d] the public interest" and "infring[ed] on a person's privacy."[58] As viewers grappled with this nuisance, they showed no credible signs of giving up on television. Instead, they reached out for reform in an effort to make television a better-behaved live-in companion.

An Unwelcome Guest in the Home

For the regulators and the industry participants who had toiled for years to erect a suitable infrastructure for nationwide television service, the act of getting receivers into homes was a final and crucial step. By the end of 1951, the number of sets in homes had increased to just under 15 million.[59] For these institutions, television service was analogous to a guest in the home. If we accept this analogy, then we can surmise that the guest's invitation was the television receiver. Guests rarely settle in for hours at a time every day at all hours of the day, however, so television became more of a roommate or a member of the family. Roommates and family members fight, so it is not surprising that some viewers expressed frustration with this contraption that insinuated itself into their lives with little respect for their values. One such viewer feared the ease with which that insinuation occurred: "Frankly, I am somewhat alarmed and fearful of letting an unguarded and uncontrolled visitor [. . .] come into my home and not only affecting the lives of my family, but possibly the thinking and the future development of those that are taking part in it."[60] Parents writing to the FCC assumed some of the blame, chastising themselves for inviting the guest into the home and exposing their children to "such demoralizing things."[61] Others preferred to distance themselves from culpability, diverting blame via the passive voice: "This is to register my disgust for the large amount of drinking and drunkenness and sexually suggestive scenes that are allowed to contaminate our homes on television." Still others saw themselves as hostages forced to sacrifice their values for their captor. One man wrote that the bars in his town forbade women from entering, but the reverse was true on television.[62] As a result, in his home, on his television set, his family "[had] to look at it, very disgusting."

Home invasion was an analogy that some viewers preferred. One woman, protesting the violence on the action-adventure series *Dick Tracy*, accused ABC of breaking and entering: "You have no right to invade our homes and pollute the air with your crime loving, immoral, indecent stuff. Get out of my home, and stop corrupting my children."[63] Along similar lines, a man incensed with NBC asked the FCC, "How much longer must we endure having these immoral people forcing themselves into the privacy of our living rooms?"[64] New Year's Eve footage of a "drunken mob" likewise "violated the homes of America."[65] Comedic performer Milton Berle, too, with his "crude" hand gestures, was "invading the privacy" of one letter-writer's home.[66]

Viewers' expectations of a well-mannered guest were thwarted with great frequency. The "harmful exhibition of liquor drinking" on the nightclub-themed program *Stork Club* "debased" the "sanctity, morality, and respectability of the American home."[67] As with the issue of television volume, some people felt that television programs' bad behavior was violating their rights as well as their homes. For one viewer, the will of a "producer [. . .] to inject such trash into the homes of millions of decent American people" collided with "the right that American families have to expect that nothing but clean, wholesome entertainment be sent into their homes."[68] Perhaps the most striking and literal interpretation of the guest analogy came in a letter sent to Arthur Godfrey, the FCC, CBS, and Ligget & Myers (Godfrey's sponsor).

> Dear Mr. Godfrey: You were sitting in my living room last Wednesday evening, talking to my wife, my twelve year old daughter and myself. In a conversation with your lady singer, you made a nasty off-color remark about a wide-open blouse. I promptly turned off the television set, so as to spare my wife and daughter any more of your coarseness and vulgarity. I think you owe them both an apology.[69]

Objectionable trends on television—filmed content, noise, bad behavior—enraged some of the people who had dedicated space in their homes to accommodate what they hoped would be an entertaining yet respectful guest. As charming a metaphor as the "guest" line may have been, it had several practical purposes: it eased the entry of a new, large technology into homes; it softened the blow to families' expenses; and it merged necessary household consumption with the excessive commercialism now made visible to a generation of radio listeners. By humanizing television service—transforming it into an amiable companion who drifted in and out of the living room—this sort of discourse attempted to forge natural relationships between viewers and content, all the while masking the industrial and regulatory negotiations making transmission possible. The viewers who complained to networks, sponsors, and the FCC had at least an inkling of what lay behind the charming facade, so they did what they could to discredit the metaphor.

Equal to the task, the FCC was armed with replies that displayed its deft handling of situations over which it had arguably little control. The extent of the FCC's powers over programming will be discussed in the next chapter, but here we can focus on how the commission wanted to portray its powers over programming—or, more accurately, its powerlessness. One such reply, sidelining the FCC's role, emphasized the importance of writing complaint letters. In one, T. J. Slowie wrote,

> It should be pointed out that the listening public has a certain responsibility to make the broadcasters and sponsors aware of the objectionable features they find in program content. Generally, the broadcast industry is influenced by expressions of opinion from the listening public, and the extent to which people communicate with the parties involved can have a particularly strong bearing on the selection and presentation of program material.[70]

In his reply to a letter about the visual and aural offensiveness of NBC's crime drama *The Big Story*, Chairman Coy maneuvered around the FCC's responsibilities with tactical care.[71] He wrote, "At the outset, we must agree that no Commission should be the arbiter of public tastes, dictating what you, I, or anybody else, with our varying tastes and interests, may or may not view or hear." He then inserted the industry into the conversation: "It is clear, too, that there is a genuine issue as to whether the broadcasting industry is fully discharging its obligations to serve the public interest." Next, he inserted the FCC to prove that he understood the problem and had tried to urge someone else to solve it: "For my part, I have taken the position, and publicly urged in various addresses before industry and other groups, that radio and television must begin immediately to clean its own house." Finally, Coy brought up the newest tool at his disposal, a tool he could employ to reroute complaints: "I am hopeful that the NARTB, which is now drafting a code to govern television broadcasting, will find the answer."

A reply from Slowie to a general complaint about NBC, written after the adoption of the Television Code, credibly wove the viewer's concerns into the feedback mechanism installed by the NARTB.[72] Exalting the viewer for his proper judgment in complaining, Slowie wrote that programming would only get better "if listeners in sufficient numbers follow your example and make known to licensees of broadcast stations and to network organizations the character of programs that are desired and those that are objectionable." He then acquainted the viewer with the Code and explained their compatibility with each other: "Your writing to NBC [. . .] is responsive to the last paragraph of the Preamble to the NARTB Television Code which reads, in part, as follows: 'viewers should be encouraged to make their criticisms and positive suggestions known to the television broadcaster.'" Replies like these did not always conclude the matter, and viewers unquestionably were frustrated with form letters, especially since the government was answerable to the people in ways that the networks were not. There is an indication, however, that the NARTB was happy with that type of reply. Thad Brown thanked FCC chair Paul Walker in July 1952 for guiding viewers to the Television Code Review Board.[73] The visibility of a watchdog, if not a strict method of enforcement, lent credibility to the Code. Regardless of the feelings of the higher-ups, viewers wrote in consistently, responding to offenses large and small and creating the basis of a conversation that would lead to greater television-specific self-regulation.

Undoing the Nation's Morality: Gender, Sex, and Crime

Major points of concern in the 1952 Television Code were education, children, decency, and religion. Each of these and more occupied the minds of pre-Code television viewers. These were also the chief issues that dominated the discourse surrounding the radio codes. The visual language of television tended to exacerbate worries, however. So, anxieties about children's exposure to visual performances of violence extended beyond crime dramas. The "foul brutality" of wrestling made several appearances: some of the more extreme letters called such broadcasts "degenerate," "pervert[ed]," "dirty," and a "disgusting and degrading spectacle."[74] One viewer even wrote to his congressman asking for an investigation into wrestling practices.[75] Another concern that crossed genres was sexuality; some viewers were made uncomfortable by the transgression of gender norms in wrestling and comedy. In her work on television regulation, Hendershot proposes that censorship is a "social process through which the politics of class, race, gender, violence, and other potentially 'problematic' issues are deconstructed and reconstructed, articulated and scotomized."[76] The normative assumptions ingrained in acts of regulation and censorship, while not central to this book, reveal themselves in the text of the Television Code itself and in the words of viewers incensed by early TV programs. A viewer criticized irreverent performers Milton Berle, Bob Hope, Dean Martin, Jerry Lewis, and Spike Jones for their "unfunny feminine antics."[77] Especially cutting was one letter, again pointing to Berle, which read, "Time Magazine says there are 5,000 homosexuals on government payrolls in Wash[ington, DC]. Let's hope FCC is not in favor of Berle and his kind."[78] The flamboyant wrestler Gorgeous George incurred similar treatment. A concerned mother demanded that the FCC "throw him off the screen if not out of the ring before we have the children wearing lamé robes, throwing flowers, and wearing long highly hued tresses."[79]

In addition to lobbying to protect sports from the excesses of wrestling and heterosexuals from the gender play of television personalities, viewers acted to shield nationhood from any hint of disrespect. According to viewers, respect for national symbols, such as the flag and the national anthem, deteriorated when displayed within the context of television's commercialism; distracted audiences ignoring these symbols also intensified their mistreatment.[80] More insulting to a greater number of viewers was political satire. "George Washington Slipped Here," a skit on an NBC variety show broadcast on the Fourth of July, prompted one woman to write, "It ridiculed our American heroes and our American institutions in a way that the Reds themselves could not have topped."[81] A like-minded individual protested an episode of another NBC variety program, *Your Show of Shows*, in which "President Truman was ridiculed."[82] Confident of FCC action, the viewer wrote, "Who was guilty and what should be done about it is your business." When comedian Red Skelton called a half-dollar a "Truman Dime," a viewer wrote to demand he be taken off the air.[83] Another insisted that "humor degrading this high office should be outlawed."[84] For these viewers, the FCC did not adequately safeguard "the respect and dignity" of elected officials.[85] Not only was television unraveling the nation's moral fiber, but it was eroding the reputations of authority figures.

". . . It's Almost a Waistline, Not a Neckline . . ."[86]

Letter-writers demanded respect for ethics, gender roles, nation, and authority. Likewise, they demanded that the government and the industry uphold a specific definition of decency and a high regard for law and order. In fact, one-third of all letters examined here focused on indecency. Of these 353 letters, 64 percent focused on costuming (mostly low necklines and strapless gowns), 14 percent focused on bawdy jokes, 8 percent made reference to sex, kissing, or sexual suggestiveness, and 6 percent characterized dance routines as burlesque.[87]

Costuming, a matter incorporated into the Television Code under "Decency and Decorum in Production," elicited a wealth of outrage, mostly from women, though men were not shy about offering their input (see appendix B).[88] One man wrote of his opposition to "spectacles of voluptuous hussies exposing large expanses of their anatomies."[89] Another found "disgusting" Lena Horne's "dress, postures, and heavings."[90] Also "disgusted" was a man who noted an incident in a CBS program in which "one of the girls bent over and the bra was not of much use."[91] Opting for proper terminology, a male viewer notified the FCC of "women on the network who wear formals that are becoming so low and wide enough to expose part of the mammary regions and almost low enough to reach the epigastrium."[92] In a prescient bit of commentary, one man, who admitted he "appreciate[d] certain things as much as the next fellow," criticized the practice of "expos[ing] breasts" because "that alone is going to be the basis upon which feminine performers are going to be hired."[93] And an entire family asked, "Why do the ladies wear dresses that make the people in decent homes ashamed to turn on their TV sets? Dress them more."[94] Some men admitted embarrassment not at the women's dress but at the men's wardrobe. Writing about the "pants worn by 'boys' in ballet dances," one male viewer asserted, "they are a scandal."[95] A group of male office coworkers composed a letter complaining about Arthur Godfrey's show and pointing specifically to Godfrey's pants: "Godfrey's costume was very evidently planned as an opportunity for him to exhibit certain personal appurtenances of which he must be proud—in a moronic way—else he would not have made such a public spectacle of himself."[96] These men were not alone in their assessment: a woman also noticed that Godfrey's pants were "tight in spots."[97]

Female letter-writers expressed shame, anger, and embarrassment at low-cut and strapless gowns. "Some of these bold girls need their faces slapped for the dresses they wear," one wrote, while another expressed her disgust with "bare women."[98] Another was convinced that audiences did "not want to see this 'gutter' work of actresses trying to outdress the other or be more daring than the other."[99] Complaining about "bosomy women," one viewer asked, "Why must women show their bosoms, certainly the housewives don't go about showing theirs!"[100] The extent of the cleavage exposure threatened to

FIGURE 4.1. "But, Boss, she violates the TV Code from EVERY angle!" (cartoon by Sid Hix, 1951)
Referencing bombshell performers like Dagmar, Sid Hix's cartoon anticipates the difficult task of adhering to the strict Television Code. Cartoon by Sid Hix for *Broadcasting-Telecasting* (November 5, 1951). Courtesy of *Broadcasting & Cable*.

defy containment: "On Arthur Godfrey's show two weeks ago, the guest stars looked as if they would burst out of their dresses."[101] At least one woman dismissed this type of reaction. She believed that "if the grown ups didn't have their minds where [they] shouldn't be, low necklines would mean nothing except maybe a beautiful body."[102] Her quarrel was not with costuming but with the suggestiveness of televised ballet.

Some letter-writers shifted their irritation at so-called nudity onto the subject of male-female relations. A woman who identified herself only as "Just a Female Who is Totally Disgusted" proclaimed that "married life [would] be a lot happier if we quit putting these naked girls before the eyes of our children and our husbands."[103] Another woman who identified herself as married wanted the FCC to "imagine what a dance of immorality would do to teenagers and those in their early twenties—unmarried—the desire it inspires."[104] Another married woman attempted to explain her opinion of strapless gowns by aligning it with possessive, heterosexual, marital norms:

Most women love and respect their husbands and have enough modesty that they wouldn't dress in this manner. I truly believe that most women feel as I do—they want to keep themselves for their husbands alone, not for the public. A wife belongs to her husband. Can you understand that for the same reason a woman doesn't want her husband to look at another woman when they are so briefly clad?[105]

Other letter-writers connected female performers' manner of dress to moral decay and even to violence. A female high school student wrote that strapless dresses were "very disgusting and sickening" and were "the very cause of so many sex killings."[106] The children—"leaders of tomorrow"—would wind up being the "leaders in sex crimes" because entertainers lacked "judgment" in their "style of gowns."[107] "Plunging necklines," for one viewer, created a "nightclub atmosphere" that was "bound to create a certain percentage of moral degenerates."[108] Women, too, were implicated in what was diagnosed as pathological nudity. One woman wrote, "I have read that fifty-three percent of our hospital patients are mental cases and I believe that the low morals of our entertainers are responsible to a certain extent. It is certainly not normal for women to have such a strong desire to expose their bodies to the public. [. . .] We will always have rape and murder so long as ideas for them are given so freely."[109] The sight of a woman's body, even a mostly covered one, was thought to damage young boys. A father wrote,

> Such stuff has no place in the American home, particularly where there are growing children—most particularly, boys. What effect does the view of half-naked women have on boys who are just beginning to feel the impulses of sex! It can destroy them both spiritually and physically. [. . .] Let us protect our children against the devastating corruption of lust. It leads them to insanity and crime.[110]

Dagmar, a performer known for her sex symbol persona, was one frequent target, as was French singer-actor Denise Darcel, whose appearance on *Arthur Godfrey and His Friends* triggered multiple letters of protest. Dagmar appeared on *Broadway Open House*, NBC's late-night variety show, and for one viewer her gowns "suggest[ed] complete failure of the cotton and flax crops and plague among the silkworms."[111] No fewer than eight letters in the Arthur Godfrey folder commented on Darcel's appearance. Not only was she apparently wearing "no more than a string of beads and a smile," according to one female viewer, but the routine she performed with actor Mary McCarty "could only have originated in a house of prostitution," one man wrote.[112] "The wenches," he continued, "were attired for a strip tease."[113]

All of these wardrobe missteps may not have signaled poor judgment or a lascivious nature; they could have been communist plots to derail US morality. Decrying a scene between Ed Wynn and Joan Blondell on *The Ed Wynn Show*, one viewer wondered if, "knowing the techniques of communist thinking—their objective the breaking down of Christian ideals and morality," the communists were "writing the script" for NBC.[114] In her complaint about Godfrey's antics, another viewer wrote, "The Communists hope to corrupt our young people and far be it from us to help them to do it."[115] The FCC needed to step in; clearly sex appeal was a matter of national security.

The experience of seeing radio shows transformed into showcases for sexy attire reminds us that these viewers were unsettled not just by strapless gowns but by television itself. One letter-writer summed up her acclimation to the new medium by comparing the shock of seeing skimpy outfits to her memory of watching motion pictures showing trains rushing at the camera.[116] While the stories of early movie viewers jumping out of their seats to avoid the oncoming train may be a colorful but exaggerated way of thinking about the experiences of these early television viewers, the phenomenological accounts offered up by viewers in relation to women's bodies reveal a particularly sexist tone that materialized in the Television Code. Indeed, with the exception of the entry on race and nationality, the entirety of the Code section entitled "Decency and Decorum in Production" fixates on anatomy, costuming, proper body positioning, and sexuality—all rules potentially violated by men but most often applying to women and the camera's penchant for fetishizing female bodies (see appendix B). It is telling, then, that the Code tried to regulate not just language but production technique, an issue detected early in 1952 by a viewer who railed against an ABC program with Don Ameche that featured nudity "grossly exaggerated by the peculiar lighting employed."[117]

". . . Fistfights Lead to Gunfights and Gunfights Lead to Murder and Murder Leads to Hanging . . ."[118]

Representations of crime were another major complaint in the surviving letters. In 1951, a National Association of Educational Broadcasters study of a week's worth of programming in New York found "2,970 'acts or threats' of violence—more than seventeen 'acts or threats' per hour during children's viewing periods."[119] In the results of a companion study, which a letter-writer read in *Scientific American*, crime dramas had "increased from an already deplorable 10 percent to 14.6 percent of the total broadcast time."[120] An exasperated mother conducted her own survey and found that in an evening's worth of programming in San Francisco, there were "thirteen murders, fourteen slug-gings, fourteen kidnappings, six hold-ups to say nothing of explosions, black-mail, thieveries, armed robberies, lynchings, and torture scenes."[121] She went on to use a common tactic of letter-writers anxious to curb crime on television: recounting newspaper stories of murderous children influenced by television. "Then you read in the daily papers where a fourteen year old boy kills his mother and burned her body. Saw such on TV, he said. Another lad, sixteen, killed (shot) his father because he dared turn off a murder TV pic."

Writers like these, mostly parents, had lost tolerance for TV because of statistics like the ones offered by the Southern California Association for Better Radio and Television (SCABRAT), which found in May 1951 that the "average child in the television home sees death inflicted by violence more than forty times every single week."[122] That study sampled the forty-nine children's television programs airing in Los Angeles, and the results discred-ited the efficacy of network policies, such as those found in NBC's *Responsi-bility: A Working Manual of NBC Program Policies* from 1948 and its 1951 *Radio and Television Broadcast Standards* manual. NBC's 1948 policy for children's programs, which applied to both radio and television, forbade "material (in-cluding sound effects)" that would "create in a child's mind morbid suspense or other harmful nervous reactions."[123] The spirit of that provision remained intact in the 1951 publication.[124]

Chairman Coy invoked SCABRAT's study in a speech to the NARTB Television Standards Committee in June 1951, and a year earlier he recounted another study by the same group in an interview with the *Washington Eve-ning Star*. According to that study, one week's worth of programming in Los Angeles yielded "ninety-one murders, seven stage holdups, three kidnappings, ten thefts, four burglaries, two cases of arson, two jail breaks, two suicides, 'and more cases of assault and battery than could be tabulated.'"[125] The situation clearly had not changed from 1950 to 1951. In 1951, of the 967 complaint letters the FCC received over an unidentified 75 days that same year, 73 centered on crime and horror (while 221 targeted "indecency, obscenity, or profanity").[126]

According to NBC's 1948 standards, "Crime and punishment are treated in a way that does not portray a criminal in an attractive light nor condone a crime. Criminals are always punished either specifically or by implication."[127] Westerns featured white-hatted heroes punishing black-hatted outlaws. Gangsters, having committed their misdeeds, met their comeuppance at the hands of the state. Yet some viewers doubted the effectiveness of that prosocial frame. Asking for a legislative solution to the problem of crime programming, one viewer wrote to his senator in 1949 and stated, "I presume the argument for this type of program is that they show 'crime does not pay.' However, it seems to me that this argument is far outweighed by the psychological disturbances and sordid mental impressions created in the minds of youngsters who, I am certain, comprise a very substantial portion of the video audience."[128] "Crime does not pay" also fell flat for another viewer, who pointed out that the concluding message did not cancel out the abundant representations of violence: "Even if they say 'crime does not pay'—the great, great mass of people do not get into crime—but kids have to see it everyday."[129]

For these viewers, the frequency with which crime and mystery programs appeared on the schedule was worsened by their time slots. One letter-writer felt that crime programming "should be banned from the air in the early hours of the evening" and another felt that these programs should not be aired "during such hours when children are likely to listen."[130] One mother refused to buy a television set until a child could "watch anything being shown at 7:30 and not have nightmares at nine."[131] Requests for crime programming to be scheduled after 9 or 10 p.m. appeared frequently.[132] A petition from a parent-teacher association to a congressman likewise requested that "all murders, killings, vicious crimes and horror programs be barred from radio and television during waking hours of our young children."[133] Titles also compounded the problem of scheduling. A viewer startled by a seemingly benign television play wrote,

> At least you are forewarned by newspaper programs when a Cowboy and Indian picture or a murder mystery is telecast. [. . .] But when a reputable Theatre of the Air announces a title such as "Double Entry," you don't expect the murderous, emotion-shattering ending that we got. As a citizen I appeal to you to take the proper action that no similar programs be shown under misleading titles—especially before children's bedtime.[134]

The scheduling problem was not foreign to NBC. The network had specific standards in place to avoid the complaints raised by these parents. NBC's 1948 manual states, "No series of crime and mystery-type programs is broadcast over NBC before 9:30 p.m. Eastern Time, 8:30 p.m. Central Time, and 9 p.m. elsewhere."[135] Moreover, crime and mystery programs required buffers of "twelve intervening [noncrime or mystery] broadcasts" so as not to bombard the viewer with darker subject matter.[136] For reasons that will be discussed later, the network's rules could not always be translated to the screen.

Letter-writers' declarations that they would remain outside of the television audience by not purchasing a set represented an extreme response to crime programs.[137] The more compelling letters lamented a fractured viewing experience. One mother wrote, "It's getting so you can't sit down with your children and enjoy an evening of T.V. I have five and it is a fight every night to keep them from watching these crime shows. The older ones want to see them and if they do watch one now and then I have to stay up all night with the little ones."[138] The focus on children detracted from the effects that crime programming had on the way adults watched television. Writing about both radio and television, another mother pointed specifically to a gendered viewing experience: "The men in my family insist upon listening to all the gruesome murder stories and do not see the danger to the children."[139] This letter highlights a splintering of viewership that seems to have occurred in prime time. Graphic content may have pushed some women away from the screen, but the advertising of female-oriented products on early prime-time crime programs, Catherine Martin argues, proved that sponsors expected to find women in the audience.[140] The appearance of "sympathetic female characters" in crime programs was another indication that the networks wanted to target women alongside men.[141]

Graphic content also prompted some letter-writers to contrast their new viewing experiences with those of radio listening. The "radio was bad enough" refrain recurred in these letters and tapped into an anxiety about the combination of seeing and hearing simultaneously.[142] For one man, television was "a great deal worse" than radio, and he recounted an evening of viewing in which "the actual scene of fighting and murder with the groans, blood and struggles were just too much" for him, a self-described "tough" guy.[143] The disturbing sensory experience prompted one viewer to wonder about the mental health of the TV writers: "We 'see' smooth, smart gunmen and gang wars, murders, a siege of terror in underground tunnels [. . .] we see peeping Toms watching a woman undressing, we hear hair raising shrieks, and finally, a floating body in a river."[144] A particularly harrowing letter directed to CBS relayed an incident in which the wife of the letter-writer became "actively ill from watching [. . .] a man being beaten to a bloody pulp in full view of the camera."[145] This, for the woman's husband, "was brutality for the sake of brutality."

NBC did not wish to encourage such fraught sensory experiences. Its 1948 policy stated, "The use of multiple crimes of violence and the use of horror for its own sake are not permitted. Brutal killings, torture or physical agony are not presented in detail, nor indicated by offensive sound effects. No character is depicted in death agonies. Sound effects, calculated to mislead, shock, or unduly alarm the listener, are not used."[146] An equivalent passage eventually appeared in the Television Code, so once again, visual and sonic expressions sat alongside linguistic ones in documents attempting to rein in perceived excesses.

Unsurprisingly, the horror genre, as it has been imagined by filmmakers, suffered from such restrictions while crime shows and mysteries continued, allegedly contained by the dominance of law and order. Scholars and creative workers generally agree that network standards and the Television Code stifled frightening aesthetics, and some have argued that the type of horror that did develop on television was inauthentic because it lacked the "graphic 'splatter' horror that [was] possible in novels and films."[147] Matt Hills examines the claims of "authentic" versus "inauthentic" horror that have placed television horror on a lower aesthetic rung than movie horror.[148] Scholars such as Gregory Waller, Lisa Schmidt, and Andrew Owens reject the claims of inauthenticity, arguing instead that television horror was a different type of horror that capitalized on enforced, "indirect" horror, on the genre's roots in melodrama, and on television's industrial and creative predilection for seriality.[149] We should also not forget that television inherited and expanded upon the type of generic hybridity evident in radio's thriller and horror anthology programs.[150] Without a doubt, some viewers were genuinely horrified by the images and sounds that seemed out of place and inappropriate in their living rooms. The legacy of early regulations is apparent in the proliferation or diminution of genres, and that speaks to a fascinating exchange between the industry, the audience, and creative labor. However, one of the material reasons for some viewers' agitation was the result of decisions unrelated to television content.

The seemingly interminable license freeze exacerbated some viewers' frustration with crime-heavy programming. Served by only a DuMont affiliate, one letter-writer implored the FCC for "some new channels" so that she and her family could "look at something besides murders + crimes."[151] The year the freeze ended, one viewer with no television service wrote to the FCC and to her town's mayor, worried that the impending presence of TV in Oregon would mean an influx of unwanted crime programs.[152] On the opposite end of the TV service spectrum, a viewer from Philadelphia, which boasted three stations before the end of the freeze, wanted the FCC to ensure "freedom of selection" by mandating that one station be "free of crime from 9 to 9:30 and 10 to 10:30 on Thursday nights."[153] So even the presence of multiple stations did not ensure the variety that some letter-writers craved.

Despite the political wrangling and handwringing over frequency allocations for educational purposes (some on the FCC felt that the proposed number of allocations was adequate, FCC commissioner and educational broadcasting advocate Freida Hennock wanted more, and the NARTB thought the number was shockingly high), the FCC was able to position new educational stations as an antidote to these viewers' complaints. In his response to a parent-teacher association in March 1952, FCC chairman Paul Walker posited that the new educational TV allocations "may well become a primary factor in the improvement of broadcast programming."[154] The FCC also hoped the NARTB's Television Code would stem the tide of letters. By 1952, the commission's reply letters included information about the Code, which was a convenient way of deflecting criticism.

The FCC wanted viewers to direct their ire to the parties responsible for content, and viewers were doing just that. Some viewers contacted the networks directly while sending courtesy copies to the FCC or complained directly to the FCC about the networks. "CBS owes its audience a public apology," one letter stated, while another claimed, "NBC is only interested in one thing: how much TV time they can sell and the devil with what happens to children's minds."[155] Some viewers also reached out to sponsors, who sometimes protested their innocence and forwarded the complaint to the network.[156] Of the entirety of the letters examined, only one revealed that a complaint had been resolved to the viewer's satisfaction. In 1951, a man complained about a murder scene in the Pall Mall-sponsored *Big Story* on NBC.[157] In addition to receiving a response—generic, no doubt—from Chairman Coy, the viewer received a letter from Pall Mall, and "on a subsequent telecast, it seems that a gory murder scene was eliminated and the announcement made at the end of the broadcast that certain portions of the film had been deleted." In her study of listener mail and the radio program *Barn Dance!* Kristine McCusker observes that letters could affect the life of a program.[158] Sponsors, especially, were "impressed" when many listeners wrote in to a show; stacks of mail "proved that advertisements could be heard and products sold."[159] As the Pall Mall example illustrates, stacks of negative mail proved that products could be boycotted as well.

Whether or not individual fixes created meaningful change, the industry finally resolved not to confront each misstep in a piecemeal fashion. As chapter 3 explained, the decision to craft and implement a code was not easily reached, but once completed, the Code sought to centralize viewer feedback via the Code Review Board, bypassing government watchdogs. With a successful Code reining it in by industrywide standards, TV would be safely out of the critical spotlight and no longer a source of controversy. The self-regulatory maneuver also sought to tamp down calls for censorship, which emanated from politicians but also saturated complaint letters. Justin Miller's rhetoric lived alongside and completely contradicted the requests by a number of viewers for government involvement, underscoring how far apart the industry's goals were from the letter-writers' desires.

"Censorship We All Abhor, But . . ."[160]

The FCC's most popular reply to letter-writers—that Section 326 of the Communications Act of 1934 prohibited the commission from censoring programs—was intended to shut down requests for censorship. One viewer responded to the FCC's reply, however, by arguing that the commission's duty "in charity or love with God and man" was to "counsel and admonish." Such was the tenor of many letters that took to heart the theme of moral decay. The persistence demonstrated by some writers who replied to replies, thus entering into a long-term conversation with the commission, illustrated their conviction that television presented a serious challenge to the nation. One writer went so far as to reply to T. J. Slowie and demand legislation that stations operate as a public service.[161] Another viewer who wrote to the FCC multiple times conceded that censorship was something "we all abhor," but insisted that something needed to be done about the crime and indecency on TV.[162] Slowie replied that the new Television Code, recently adopted by the NARTB, might obviate censorship and still satisfy concerned viewers.[163] The viewer to whom Slowie replied was not alone in his ambivalence about censorship. A parent who had to resort to forbidding his preteen son to watch NBC's *Leave It to the Girls*, a female-centric talk show, wrote, "Personally, I am opposed to television censorship, but if it would mean control of programs of this type, then I'm for it."[164] A woman recounting the fear stirred by a horror program wrote, "We certainly are against censorship in our daily life—but we do feel very strongly that there should be censorship of evil and crime and 'scary' stories such as the one on the air last night."[165] Viewers like these protested the idea of censorship presumably when it came to serious things like news or literature, but when they found their home life inconvenienced by what the industry had constructed as a leisurely pursuit, then the First Amendment was up for debate.

One viewer was "shocked" that radio was censored but television was not; he believed that television "require[d] it far more than radio."[166] He may have been thinking of the NAB's 1948 radio code, which was not government censorship, but by appealing to the FCC he was asking for greater government involvement. The presence of censorship boards for motion pictures but not for television confused one viewer.[167] In his letter to the four television networks, one writer offered the companies a choice: either a "voluntary board or full time committee of educators and others to pass on the type of entertainment that you and your advertisers attempt to broadcast into our homes," or a government-appointed "board of censorship that would be objectionable to everyone, in giving the Government too much influence on our thoughts, and will hamper your company in the free operations of your business."[168]

Other viewers thought a censorship board for TV already existed. One wrote that he believed such a board had "been lax in permitting passionate, long-kissing to go on unmolested in television shows."[169] Another expressed his assumption that the FCC had "more or less control of programming."[170] The assumption that the FCC could "allow" broadcasters to air certain programs was also prevalent.[171] A clergyman wrote, "I believe it is in your power to curb many things that are distasteful to the Christian heart and mind that are transmitted via the radio and tele-vision."[172] For another member of the clergy, without government censorship the public was "completely at the mercy of these TV stations."[173] One viewer invoked the proper keywords in his appeal to the FCC: "With your authority of regulating licenses you owe it to public interest to do something about this horrible situation."[174] One parent-teacher association even wanted parents to constitute the censorship body.[175]

The majority of viewers asked the FCC point-blank to do something. One particular viewer, upset that the money spent on the set was wasted because she constantly had to turn it off to avoid "sickening programs," wrote, "There is something that can be done about it, and we expect you 'Men' to do it. Are you 'Men,' will you do it? Well we shall wait and see!"[176] A teenager joined the chorus: "In giving the equivalent of a censor board for television you will be doing three things: 1. fulfilling your duty to the people, 2. properly using the necessary authority invested in you, 3. making America a better place to live."[177] Aware that the FCC's charge was tied to the public interest, several letter-writers echoed this theme of duty.[178]

Not everyone was keen to have a censorship body. One viewer wrote specifically to denounce the attempts by the Catholic organization Knights of Columbus to encourage government censorship of television.[179] The Knights of Columbus had recently and ambitiously "demand[ed] that [the 'proper authorities'] take the necessary action to see that all television shows are presented in such a way that they will not offend any person."[180] An announcement that the FCC was proposing an inquiry into television programming elicited more anti-censorship responses. One viewer refused to label programs as "wholesome," but he did not believe their content required government intervention. For this man, "the legal, coercive imposition of standards or rules of conduct can at best be only a half-informed attempt to make everyone conform to the moral values of a (usually vocal) few."[181] For one father, the "crime does not pay" message was a factor in the acceptability of crime programs, which his children "enjoy[ed] observing."[182] Furthermore, he argued, no link existed between these programs and crime. Cars—not TV—were actually

responsible for deaths, and "girls show[ed] more [. . .] at the beach." He concluded by reminding Coy that "we don't live in the year of 1915." Such a progressive view was rare among the letter-writers, but not all anti-censorship
letters were tactful. Well after the Code was put into practice, one man pled
with the FCC "to preserve what little liberty we have left in this country by re-
fusing to let frigid women, inhibited males, and bigoted clerics clamp censorship on TV."[183] He concluded by lamenting the speed with which the industry
"put clothes on Faye Emerson." One Milwaukee woman, barely disguising her
stake in the brewing industry, argued against censorship on the grounds that
it would unduly affect programs sponsored by beer companies.[184] "Prohibition
groups [did] not like programs fostered by breweries," she wrote, but brewery
workers were "among the highest contributors to the internal revenue department." Her solution to objectionable programs was a common one among
anti-censorship letter-writers: "turn off the set."

Other viewers saw censorship as the punishment for negligent self-
regulation. Arthur Godfrey received one such admonition. In a letter written
directly to Godfrey, a father predicted that the chorus of objections to the
comedian's "coarseness [. . .] will be instrumental in bringing censorship to the
whole television industry." He continued, "It seems a shame that a few people
like yourself must expose a great business to that threat."[185] Affronted by the
ABC broadcast "Reunion in Vienna," one man warned, "It only takes a few
faux pas by low bred producers and VPs to cause strict government censorship,
which we all dislike but obviously need after last night's orgy."[186] This particular episode of the short-lived anthology series *Celanese Theatre* incensed a
number of viewers because of its sexual themes. Another outraged male viewer
wrote that the television industry needed to examine material like "Reunion"
more thoroughly if it "wish[ed] to avoid direct censorship of all material by the
Federal Government."[187] Writing directly to the sponsors, a man representing
a Catholic organization reminded the recipients that "it was this sort of thing
that brought censorship to the field of motion pictures."[188]

Pleas for censorship were incompatible with the NAB's crusade to paint
government regulation as a threat to free enterprise and democracy. Where
the NAB and some of the letter-writers found common ground was in the call
for more aggressive self-regulation. That option was rarely considered in the
letters, however; the vast majority of viewers who wrote to the FCC and mentioned censorship wanted the government to take charge. In all, sixty-three of
the surviving letters explicitly requested government censorship, which was
sixty-three too many if we believe that television broadcasts are protected
speech. Certainly, Justin Miller would not have been pleased. We see, then,
how the institutional rhetoric of the NAB—censorship destroys commercial

and democratic ideals—and the personal rebukes of viewers—the products of an unchecked commercial system endanger the family unit—closed in on both the FCC and elected officials, creating a palpable tension within the government and leaving a generally contentious atmosphere in their wake.

Viewer Responses to the Television Code

The adoption of the Television Code in late 1951 and its implementation in March 1952 gave the FCC an extra paragraph in its reply letters and an additional escape route to accompany Section 326. The FCC was clearly eager to deploy this new weapon to deflect demands for government censorship, but less clear was the Code's success as a public relations weapon. It must be noted first that some letter-writers pushed for a code and not for outright government censorship. The Methodist Radio and Film Commission, for example, wanted the FCC to pressure the industry to adopt a code.[189] Upon hearing of the impending Code, some viewers were pleased, although one was certain that the FCC (and not the NARTB) would enforce it.[190] As we can see, confusion about the line between government regulation and industrial self-regulation commingled with confusion about the commission's job in general. The FCC tried to make that clear in at least one reply: "The code is the result of the voluntary action of [NARTB], this Commission having had no official part in the adoption of it."[191]

Reactions to the Code were not overwhelmingly positive. Before the Code even went into effect in March 1952, one viewer held out little hope for its efficacy. "When the TV industry tell us about their self-enforced code, how strict it is, I get a pain in the belly," he remarked in his letter.[192] Early in the Code's life, few people knew of its existence, much less its components, so the FCC's paragraph about the Code had educational value. One viewer who responded to the FCC's reply appreciated the news about the Code: "Quite a few of the Los Angeles stations boast of it, and I didn't know its standards."[193] The information did not convince her of the Code's effectiveness, however. "To me the NARTB is a farce now that your office has informed me of its standards," she wrote and proceeded to list several programs that violated the Code. Another viewer (one of the FCC's repeat complainers) protested against stations in Syracuse and Utica, New York, for showing movies "that condone divorce, suicide, immoral actions, etc."[194] Sharing the Los Angeles letter-writer's disappointment, he continued, "The NARTB code seems meaningless as they do not enforce the regulations." And three months after the Code's implementation, a father whose disabled son depended on the companionship of the

FIGURE 4.2. NARTB Seal of Good Practice (1952)
The Seal of Good Practice represented a station's commitment to responsible and wholesome television. Courtesy of National Association of Broadcasters.

TV relayed his disappointment in the deficient enforcement of standards. "I thought there was a code of decency or ethics that the stations were using," he wrote. "It seemed to me that [CBS's sitcom *My Little Margie*] could stand a little application of the code." This man's circumstances add another dimension to the Code's enforcement. As the father of a boy whose movement was limited, he did not want to police his son's enjoyment; he depended on the industry to do that.

One man, eager for television to cancel all singing, dancing, and films, wrote, "There is a seal that each of the television stations carries, it is the NARTB, there are four things written on the Seal [entertainment, education, culture, and information]. Television has given none of these to the American people."[195] The chairman of the American Bar Association, criticizing a television station for showing a true crime program, asked, "Of what value is a code of television conduct when it is flouted with such indifference?"[196] He specifically referenced the Code provision that "programs should avoid detailed presentation of brutal killings and torture." The program, as he described it, consisted of a dramatization of a murder and actual footage, taken by the

St. Louis Police Department, of the accused killer "detailing each step in the degrading crime." The failure of the Code to prevent the airing of this local program revealed cracks in the enforcement mechanism. Some letter-writers were unconvinced that the Code even existed. Incensed at a liquor-infused performance by comedian Joe E. Lewis on the CBS variety show *Toast of the Town*, one man asked, "Where is this high sounding code that was entered into by the radio and TV interests?"[197]

The Seal of Good Practice, elaborated on in the conclusion of this book, promised something viewers were not seeing, hence the bewilderment that some of these letters evince. The scope of this book precludes an examination of post-1952 complaint letters. However, if the letters addressing the Code immediately after its implementation are any indication, the tenor of the subsequent years' letters cannot have changed in any meaningful way. If the Code accomplished anything substantial at the outset, it gave viewers a more clearly defined outlet for their complaints and a tangible document to accuse of failure.

The TV Complex Responds

As early as 1949, NBC executives had complaints about television on their radar. Walter J. Damm of the NBC affiliate WTMJ wrote to NBC vice president Pat Weaver in November expressing concern about the network's "seemingly loose supervision of television mystery shows."[198] His proposals were to shift *Fireside Theater*, *Big Story*, and *The Clock* to later time slots (even if he had to air kinescopes) and to request descriptions in advance. For Damm, the problem was that NBC was "failing to realize that what metropolitan New York [. . .] will accept is not acceptable in such towns as Milwaukee, St. Louis, Cincinnati, etc." Damm also pointed to the difference between horror on radio and horror on TV, confessing that the shows gave even him nightmares. Referencing the circumstances of the license freeze that the viewers suffered through, he wrote, "We are a one station city. [. . .] Those who are not in favor of it have no other place to go." Not long after, George M. Burbach, the general manager of the NBC affiliate KSD-TV in St. Louis, sent a letter to Weaver attaching two press clippings and two complaint letters "to add to the previous ones" addressing crime programs in general and *Lights Out*, *The Clock*, and *Life of Riley* specifically. Burbach predicted that it would "be only a question of time until public opinion will force some type of supervision if the present trend is continued."[199] Bob Hanna of WRGB in Schenectady also voiced his discomfort with "the horror and salaciousness" of *The Clock*

and threatened to drop it altogether.[200] Under fire from their own audiences, some stations—at least the larger, more established ones—were not shy about kicking the problem upstairs. A memo from Syd Eiges, NBC vice president in charge of press, to Pat Weaver written one day after Burbach sent his letter disclosed that other groups were condemning TV crime publicly:

> You may have noticed that, as was predicted at the staff meeting, protests against crime and mystery on television continue to grow, the latest and most serious being the formal protests filed with the FCC by the Southern California Association for Better Radio and Television against six of the seven Los Angeles stations. Maybe we should discuss this some more.[201]

NBC's executives mobilized, and the response was swift. Stockton Helffrich, NBC's censor, fired off a confidential memo to Fred Wile Jr., NBC's head of production, contextualizing the problem by focusing on NBC's tenuous control over its affiliates and its schedule.[202] Helffrich suggested that the network proceed with caution; acceding to the demands of stations (and, by extension, their viewers) would amount to the network's admission of fault in the larger conversation about "audience reactions to media where even the specialists (pollsters, psychiatrists, educators) [were] violently divided." He then relayed NBC's botched crime and mystery policy for radio—partially a scheduling policy (discussed earlier) that wound up hurting the network in its daytime and Saturday morning lineups. Even its prestige dramas felt the hit as NBC "found itself hard pressed to make distinctions between permissible realism (read 'horror') in productions [. . .] as contrasted with purely sensational realism in whodunits." The scheduling policy that was initially an effective public relations tool met with resistance from advertising agencies. Helffrich explained that, faced with agency noncompliance, the network "by March 1949 was considering whether to put in print the relaxation of a policy quite literally abandoned in fact three months earlier." The narrative Helffrich spun placed viewer complaints in the context of advertising agencies' power over commercial broadcasting. The NBC publications cited earlier set stringent standards for crime programming, but the agencies' refusal to adhere to them resulted in the excesses that rankled letter-writers. Helffrich's admission that agencies loved whodunits because they were "highly profitable and cheap to produce" explained their frequency on the schedule, which only yielded greater viewer aggravation.

Station aggravation was also problematic for the networks. Troubled by Walter J. Damm's request for advance copies of program descriptions, Weaver forwarded Damm's letter to the Sales, Production, and Station Relations Departments, asking, "What do we do to a revolting affiliate, who has a point but not a base for revolution?"[203] A memo from Fred Wile Jr. laid out a multipoint plan for handling station complaints, which discounted the possibility of sending Damm information about the episodes in advance.[204] Doing so "would be the beginning of 'shared control' of NBC editorial policy, and shared control is no control," the memo declared. The network would acquiesce to the kinescope request, however, because Damm's license was on the line. Additionally, the Continuity Acceptance Department would attend rehearsals of horror-related or "otherwise distasteful" scenes, but producers would bear ultimate responsibility for what made it to air. Finally, the network needed a "practical policy" for television, which apparently was not being served very well by the 1948 manual *Responsibility*.

In December, a memo responding to a story in *Variety* about crime and mystery programming suggested adapting *Dragnet* to television.[205] According to Sam Kaufman of NBC's Press and Information Department, a move like that would be a "partial answer to the squawks of NBC-TV affiliates." Helffrich loved the idea; *Dragnet*, as he put it, had "class" and "artistic stature."[206] Programming a staid cop show like *Dragnet* did little to quell viewer discontent, as

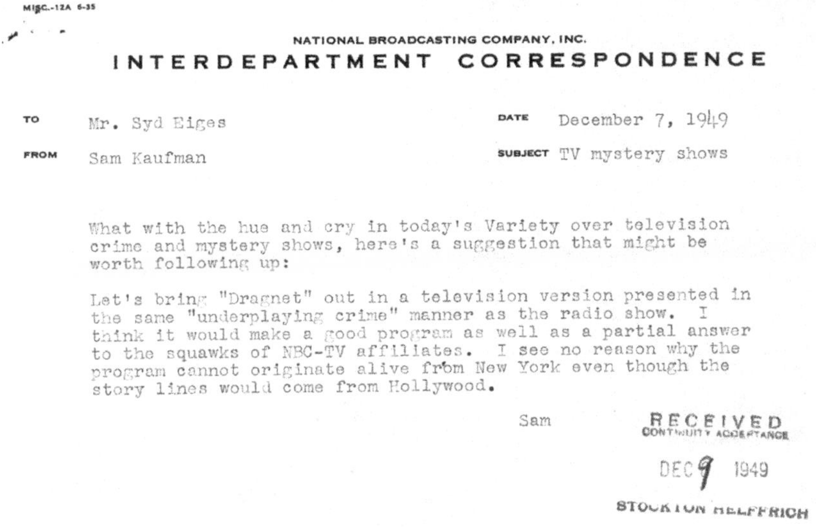

MISC.-12A 6-35

NATIONAL BROADCASTING COMPANY, INC.

INTERDEPARTMENT CORRESPONDENCE

TO   Mr. Syd Eiges   DATE   December 7, 1949

FROM   Sam Kaufman   SUBJECT TV mystery shows

What with the hue and cry in today's Variety over television crime and mystery shows, here's a suggestion that might be worth following up:

Let's bring "Dragnet" out in a television version presented in the same "underplaying crime" manner as the radio show. I think it would make a good program as well as a partial answer to the squawks of NBC-TV affiliates. I see no reason why the program cannot originate alive from New York even though the story lines would come from Hollywood.

                                            Sam   RECEIVED
                                                   CONTINUITY ACCEPTANCE

                                                      DEC 9 1949

                                            STOCKTON HELFFRICH

FIGURE 4.3. The NBC *Dragnet* Memo (1949)
This 1949 memo from Sam Kaufman to Syd Eiges proposes that NBC answer the outcry over violence on TV by producing *Dragnet* as a live television program. Courtesy of Wisconsin Historical Society, WHS-133407. Image courtesy of NBCUniversal Media, LLC.

the letters quoted in this chapter make clear. A year after the Damm affair had passed, Weaver admitted that the issue of "taste" on television "will never be solved to general satisfaction."[207] The post–Television Code letters signaled as much. NAB president Justin Miller, also apparently exhausted by the mounting viewer criticism, remarked, "It would be interesting, indeed, if the 'viewers' and 'listeners' of radio and television, who enjoy the daily programs [. . .] were as vocal in their appreciation as the zealots and reformers, who are busily engaged imposing their standards of taste upon the public, are in their criticism."[208]

Several months after the Code went into effect in 1952, Congress cast its gaze on TV. In his testimony before the House Interstate and Foreign Commerce Committee's Special Subcommittee Investigating Radio and Television Programs, Ralph Hardy, the NARTB's director of government relations, effectively dismissed the views of the thousands of people from all over the country who had written stations, networks, and the government to complain.

> I speak of the inflation of protests against this or that radio or television program which are the result of group pressure activity and not a genuine reaction of a bona fide listener. Believe me, over the years, I have talked on the telephone with hundreds of people who wrote in criticisms against a particular program, only to discover that they personally had never listened to the program in question, but were taking their action at the request of a chairman of this or that committee of a club to which they belonged. If broadcasters ever appear to be skeptics about so-called mass protests against this or that type of program, it is because they have to engage in such a careful sifting process to get to the facts in a particular case. The average broadcasting station does not receive any great volume of critical mail during a typical year.[209]

Hardy's denial of these letters' legitimacy is unsurprising, given the NAB's tendency to circle the wagons when testifying before the government. But the content problem had prompted internal action, as evidenced by the NAB's outreach to stations and networks two years earlier to determine the status of their program policies. That same problem prompted NBC to state that the crime show situation had improved between January and March 1950, a statement the viewers' letters contradict. Additionally, the NAB had expressly acknowledged the problem once it absorbed the TBA and moved swiftly to a code. Finally, that problem had moved FCC chairman Coy, in 1951, to use evidence of TV's effects on children to convince the NARTB of the seriousness of the "ten bills before Congress dealing with television programming and practices."[210] Everyone involved knew the extent to which TV content had become a trending topic, but the internal and external reactions reveal both moments of alliance and telling instances of division. Hardy's denial of the letter-writers' integrity openly pit the trade association against the viewers at home.

Conclusion

A foundational conundrum for all involved was that television's primary placement—in the home—saddled it with the family unit. Justin Miller even recognized that the connotations of domesticity weakened television. As one viewer noted, "Television programming must take into account the fact that it is basically family entertainment which should immediately bar certain types of (entertainment?)."[211] As Spigel writes, in the late 1940s and early 1950s, television was "a central figure in representations of family relationships," but that centrality, while symbolizing "unity," also demanded "division."[212] She argues that TV "was the great family minstrel that promised to bring Mom, Dad, and the kids together; at the same time, it had to be carefully controlled so that it harmonized with the separate gender roles and social functions of individual family members."[213] The letters quoted here make plain the gendered divisions that met, interacted with, and conflicted with television content.

The way the industry had positioned television made programs even more vulnerable to viewer criticism than motion pictures. The guest in the home rapidly turned into a hostile presence for enough people to disquiet both the industry and the government. The "spontaneous" letters the FCC spoke of, though not part of an organized campaign, were a mobilization of sorts. People who subscribed to a very specific definition of decency sought representation within their government. As problematic as their beliefs may have been, they performed their citizenship and expected to be heard.

The organized appeals were notable as well. The Knights of Columbus was just one organization that contacted the FCC to demand change.[214] Women's groups in particular formed a visible front against objectionable programming. Some of those found in the FCC's papers include the General Federation of Women's Clubs, the American Association of University Women, the Lafayette Mothers Committee on Mass Communication, the National Council of Jewish Women, the Mothercraft Club, the Council of Catholic Women of Chicago, the St. Louis Archdiocesan Council of Catholic Women, the Women's Alliance of St. Catharine of Siena Church, and the Sanilac County Federation of Women's Clubs. Although women are shown to occupy few positions in the industrial history of television, their domination of complaint letters and their willingness to organize in favor of television reform locate them in the middle of significant industrial upheaval.

Jennifer Hyland Wang provides important context for the presence of women's clubs in letters like those analyzed in this chapter. "Clubwomen" were essential to the radio industry's pursuit of both content and legitimacy.[215] Not only could the industry tap the clubs for daytime content, but it could also look to the clubs for well-connected groups of middle- and upper-class consumers. Perhaps most important for the purposes of this book was the clubwomen's public relations function. Wang writes that the networks turned to the clubs when they needed public service credibility.[216] An eventual rupture in the relationship between the networks and the clubwomen explains some of the antagonism evident in the television-era letters. As the 1930s progressed and the radio industry chased greater numbers of listeners, the clubwomen's higher-end status lost value.[217] Devalued as audience members, and with their role as "programming 'helpers'" upended, the clubwomen nevertheless demonstrated formidable strength as activists. Women's clubs angry about the quality of children's programming began "insisting upon content changes, threatening sponsor boycotts, and demanding a hand in program development."[218] Having endured the radio industry's ire in the 1930s, clubwomen were prepared to confront the television industry in the 1940s and 1950s.

Despite the anger and frustration that the clubwomen writers quoted in this chapter must have felt, their letters did not fall on deaf ears. As mentioned, the FCC did reply to the letters and did, in some cases, request more information from writers about specific infractions. The FCC also forwarded complaints to networks and stations, encouraging corrective action. And although only a fraction of the original letters remain in the FCC's papers, there was more than bureaucratic habit behind the efficiency with which the commission maintained and organized the letters. When the House began investigating radio and television programs in 1952, Representative Oren Harris (D-AR) contacted Chairman Paul Walker to request all complaint letters from the previous three years.[219] Complaints incited a moderate panic that led to conversations about editorial control, advertising agency and sponsor control, government control, and the moral responsibility of television in the US home. These conversations, in turn, led to a so-called solution attempted many times before in the radio industry. For the letter-writers, content was the crux of the issue. For the industry, the crux was commerce. Operating in between these antagonists was the FCC, whose arduous situation will be explored in the next chapter.

CHAPTER 5

The Federal Communications Commission

Impotent Bureaucrats, Underhanded Censors,
or Exasperated Intermediaries?

The story thus far, as told from the perspectives of the National Association of Broadcasters and television viewers, faults the government and particularly the Federal Communications Commission for failing to live up to its obligations. To hear the NAB tell it, the FCC operated as a totalitarian enterprise, leveling threats against the businessmen who were putting both democracy and capitalism into practice. From the viewers' perspective, the FCC was not wielding enough power. Both of these outlooks are shockingly simple. As we delve into the government's side of the story—first the FCC's and then Senator William Benton's—we should be open to clashing perspectives and complicated explanations. We should also be careful not to erect too tall a barrier between the industry and its government overseers. The relationship between regulators, trade associations, and networks was not always as harmonious as McChesney suggests.[1] To conceive of the mercurial relationships between Code players we need to look to competing theories of regulation, as well as to practical matters like the politics and machinations of agency appointees.

Streeter asserts that law itself is a "set of lived social relations."[2] Law begat broadcasting, Streeter writes, and approaching television in this way—as a "set of legal relationships"—allows us to see the medium "as a kind of social philosophy in practice, as a strategic enactment of ideals, hopes, and values."[3] Streeter's work was instrumental in forging new ways of studying television's social significance, but his point about law and social relations was not new. In his *Report on Regulatory Agencies to the President-Elect* in 1960, James Landis, too, underscores the social aspects of regulation. He observes that the government is "a government of laws by men," a point later taken up by Erwin Krasnow, Lawrence Longley, and Herbert Terry, who argue that the phrase "government regulation" obscures the process of people working with or against each other to perform a job.[4] They write that the "essence of the politics of broadcasting regulation lies in the complex interactions among diverse participants, not only in their day-to-day confrontations, but also in the more enduring adjustments they make in their relationships."[5] The complexity of these interactions can be driven by the seemingly solid but ultimately fluid boundary between public and private.

Jody Freeman offers a model of administrative law that complements Streeter's project and illuminates the public-private tension.[6] In her assessment, private and public are not distinct realms. Rather, private and public bodies should be understood as being in constant negotiation with each other. Government agencies are not anonymous institutions but, rather, sets of social relationships within the agency and between the agency and private actors. Clarke argues that regulation is a "political process" that necessarily involves a mix of public and private input and alliances between government regulators and private regulators.[7] The risk created by an overbearing or "overformal" regulator is an obstructionist and retaliatory industry, which is why Clarke maintains that the regulator must be cognizant of the benefits of working together.[8]

Corporate liberalism depends on the maintenance of productive bonds between government and corporations, but maintaining those bonds is a turbulent process. This chapter will discuss the history of regulatory agencies and theories of regulation in order to parse the FCC's formal and informal interactions with the NAB and to probe its more specific engagement with television content. Three distinct but interrelated images of the FCC emerge in this analysis. In some instances, the FCC is a legally dubious agency seeking to guide content and influence the industry's self-regulation. In other situations, it is a cautious peacekeeper, soothing flare-ups between the public, Congress, and the industry. And more frequently, it is an underfunded, weak (relative to the industry) champion of the public interest *to the extent that the public interest can coexist with the commercial industry*. By focusing on the FCC during the run-up to the adoption of the Television Code, this chapter aims to clarify the political and interpersonal negotiations visible throughout this book.

Regulatory Agencies: Wedged in between Congress and Industry since 1887

Although federal and state governments had aided private business through land grants, funding, and other "promotional efforts" since the early 1800s, Congress's creation of the Interstate Commerce Commission (ICC) in 1887 to regulate the railroads' behavior marked the beginning of what Robert Rabin calls a "distinctly American style of regulation."[9] In other words, government had intervened before on behalf of business, but locating that intervention in the body of a federal regulatory commission marked "a point of departure."[10] Robert Britt Horwitz, too, describes regulation as a "new and modern" means of state coercion.[11] Rabin is careful to point out that, although the Commerce Act, which created the ICC, was the first of its kind, it was not a calculated first step in a bold, long-term plan. Rather, it "addressed a discrete set of immediately pressing problems in an equivocal fashion."[12] The Commerce Act, like the Radio Act and the Communications Act that would follow in the next century, was the solution that most decision-makers agreed upon at the time.

The Progressive Era saw the passage of laws that were widely viewed as successes but that did not push the regulatory function as far as it could go, according to Rabin.[13] The Pure Food and Drugs Act, the Meat Inspection Act, and the Trade Commission Act exhibited the same problem inherent in the Commerce Act. None was bolstered by a coherent reform movement; rather, each answered a specific complaint with limited government oversight. Herbert Hoover, who would go on to regulate radio as the secretary of commerce, provided what Rabin asserts was "the single most influential and coherent regulatory philosophy between the Progressive era and the New Deal."[14] For Hoover, government could be a decisive actor in the success of businesses if they worked together. In believing the government could be (and should be) a cheerleader for industry, Hoover advanced "an associational model of public-private interaction" rather than a "policing model" of the type instituted in the Populist Era.[15] The associational model guided Hoover at the Department of Commerce (DOC), and it followed radio from the DOC to the Federal Radio Commission (FRC), established in 1927. Complementing Hoover's belief in government support of private enterprise was the New Deal pattern of implementing regulation to follow technological change.[16] But surpassing Hoover's associational model was the type of omnipresent regulation that came to characterize the New Deal—regulation that "eliminate[d] the economic uncertainties" of the marketplace.[17]

Rabin and others agree that no one has been able to determine the proper scope or limits of regulatory agencies. This ontological fuzziness owes to what Freeman calls the agencies' "dubious constitutional lineage."[18] Agencies were not part of the constitutional design and have been created by Congress to operate "as almost autonomous centers of legislative, judicial, and administrative power," according to Glen Robinson.[19] That may sound insidious, but as former FCC commissioner Nicholas Johnson argues, the original intention of the regulatory agency was to act as "an independent sentinel guarding the consumer's civil rights."[20] By assuming such sweeping powers, agencies could secure the public interest. Rabin's history of regulation complicates Johnson's optimistic explanation, but that does not erase the fact that the public interest is a vital part of the agencies' mandates. However, what could be a clear-cut commitment to the public interest has been muddied by vague mandates that allow "ill-defined" powers to multiply.[21] Such powers do not subside, as Landis points out, but, rather, "expand as more and more complex problems arise."[22] Only an agency-specific definition of the public interest directs the body as it navigates those new problems.[23]

Contemporary literature on regulation invites us to reflect on the complicated terrain that both regulators and industries must tread. Robert Baldwin, Martin Cave, and Martin Lodge describe regulation not just as a "red light," or restrictive, action but also as a "green light," or "facilitative," process.[24] They use the example of regulated airwaves as a positive action, since the alternative, a lack of regulation, would result in the type of reception interference that characterized radio's earliest days.[25] This allusion to probable market failure is paramount because market failure is the most common motivation for government intervention. Left unchecked, the market would "fail to produce behavior or results in accordance with the public interest," so regulation is required to safeguard against such failure.[26] Tony Prosser argues that the market-failure rationale cannot account fully for the breadth of regulation's functions.[27] He posits a "social solidarity" rationale as an alternative and uses the social theory of Emile Durkheim to support his argument.[28] Briefly, Prosser proposes that regulation should not always be seen as "second best to market allocation"—in other words, as a step that is never the best step to take *first*.[29] He therefore adopts a definition of regulation that, because of its breadth, can accommodate both self-regulation and regulatory goals not guided purely by economic motives.[30] If we conceptualize regulation narrowly as a rule-making enterprise, then it follows that these rules will be "seen as constraints on the freedom of business to compete in open markets" and will need "to be minimized."[31] But by shifting our outlook on the potential of regulation via the application of social theory, we can pair or supplant economic concerns with the pursuit of social solidarity. This may be the key to constructing a type of communications regulation that takes the public interest seriously. Prosser concludes by admitting the "fanciful" tone of his proposals, but he protests—and rightfully so—that the attention to social solidarity and equal access to services does reveal itself in regulatory practice.[32]

While Prosser's examples are particular to the United Kingdom, we do not need to look far in the US context to see how notions of the public good have been incorporated into broadcast regulation. Even something as technical as image quality standards protects the public as consumers (from purchases of subpar receivers) *and* as citizens (from unequal access to informational programming or emergency alerts). Far from being fanciful, language about the public good can be found throughout the FCC's speeches, and that language circulates around television content and the NARTB's Television Code.

The Broad, Inscrutable Power of the FCC

The Communications Act of 1934 gave the FCC its mandate, and that mandate has been characterized as "generous."[33] Preserving the core of the Radio Act, the Communications Act expanded the FCC's authority to include telephone and telegraph.[34] The sweeping authority granted to the FCC by the Communications Act mostly resulted from the inability of lawmakers to handle communications technologies on their own.[35] The regulatory buck was passed. So, too, were the details, such as whether or not broadcast networks fell under the FCC's jurisdiction, or how exactly the public interest applied to broadcasting.[36] For Krasnow, Longley, and Terry, the public interest mandate, though "elusive," granted the FRC and then the FCC flexibility to adapt to changing circumstances.[37] They concede, however, that what was intended to be liberating may have wound up being limiting.[38] One could argue that, by sidestepping specifics, Congress avoided creating high expectations for the commission. Adopting a similarly negative view, Robinson sees the public interest mandate as a "transfer of legislative power" from Congress to an extra-constitutional body that becomes yet another agency vulnerable to aggressive lobbyists.[39] That vulnerability, coupled with the fact that commissioners are appointed and not elected, has compromised agency accountability.[40]

The distinctiveness of broadcast regulation is regularly cited in the literature in two regards. First, scholars characterize the FCC's powers as unique among regulatory agencies.[41] Radio broadcasters sought out regulation to resolve interference issues that disrupted not only radio reception but a commercial-friendly system of broadcasting.[42] Radio regulation is the first instance of what Theodore Lowi describes as "the most advanced development in administrative process"—the federal manipulation of "whole" markets.[43] Here Lowi is referring to the practice of licensing, which both the FRC and FCC were given the power to do. Unlike other agencies, the FCC was not granted the power to regulate the prices that the industry could charge or the profits the industry earned, so the bulk of the FCC's power in broadcasting was (and is) situated in its authority to license broadcasters and to pursue antitrust violations.[44] The legal basis for broadcasting regulation, which was confirmed, clarified, and even extended by *National Broadcasting Company v.*

United States (1943), allowed for a degree of government interference unheard of in newspapers or motion pictures.[45]

The FCC's freedom, upheld by the Supreme Court, was an open wound for the NAB. In 1966, John F. Dille Jr., chairman of the board of the NAB, wrote an article excoriating what he called "government control of broadcasting."[46] Dille described a scenario in which broadcasters were forced to bend to the will of the FCC in license applications and to compromise their own programming missions to correspond to the FCC's arbitrary definition of the public interest. Faced with losing their livelihood, licensees followed orders and in doing so established a "dangerous precedent."[47] Dille's claims of government overreach spoke to the FCC's peculiar type and manner of influence.

The second way in which the literature explains the distinctiveness of broadcast regulation is by pointing to the uniqueness of the *circumstances* of broadcasting. These circumstances include spectrum scarcity (only so many stations can broadcast at one time in one geographic area), public ownership of the airwaves, the "privilege" of broadcasting in exchange for "serv[ing] the public," and the "uniquely influential and powerful" reach of radio and TV.[48] Robinson dismisses their validity, arguing that they are ideological rather than rational.[49] Valid or not, those four circumstances were strong enough to launch a regulatory apparatus based on what Lowi calls a "whole-system concept with absolute control over entry, prescription of certain behavior in the public interest, and meticulous prescription of territory and range through control of frequencies and transmission strength."[50] More than simply controlling goods or commerce, the FRC and then the FCC represent for Lowi the height of regulatory control—a type of control that eventually extended beyond terrestrial communication.[51] This is a far cry from the limited origins of federal regulation as outlined by Rabin, and it requires us to consider how the FCC—an ostensibly powerful independent commission—functions alongside the industry it regulates.

Theories of Regulation: Some Explanations for the FCC's Bad Reputation

In *The Irony of Regulatory Reform*, Horwitz cautions us to separate a regulatory agency's origin, or "creation," from its operation, which changes over time.[52] Agencies are dynamic, and often they are criticized for betraying their original missions. Horwitz's discussion of five theories of regulation can help us understand the criticisms leveled at the FCC. In public interest theory, the regulatory agency embodies "the spirit of democratic reform" in its charge to resolve "the conflict between private corporations and the general public."[53] Populist and Progressive Era government interventions, as explained earlier, constitute the two phases of this theory. One problem with this theory is what Horwitz calls its "Progressivist gloss"—its tendency to ignore the role that regulation plays in protecting industry and stimulating commercial success.[54]

Regulatory failure theory, or "perverted" public interest theory, connects an agency's origin to the preservation of the public interest but, by analyzing the agency's outcomes, posits that the agency has failed at upholding the public interest directive.[55] Failure is linked to the regulators'"overidentification"with the regulated industry.[56] It is worth lingering on this idea of overidentification to help us begin to examine the FCC's treatment of and interactions with the NAB and distinguish regulatory failure from regulatory capture. The two explanations for overidentification are "instrumental" and "structural."[57] The instrumental reasons tend to be social in nature: personnel can move back and forth between the industry and the agency, Congress can engage in bullying behavior, and administrators can maintain healthy relationships with industry professionals.[58] Structural reasons tap into the institutional constraints of the agencies and the vast resources of the industries. A government agency cannot compete with the financial might of the radio and television industries, which, through political influence, can conceivably recommend and approve agency appointees. Three other facets of the structural argument—the reliance of agencies on industry for research, the reliance of agencies on Congress for funding, and the reliance of agencies on industry for an agenda—exacerbate the disadvantaged state of the agencies and potentially set them up for failure.

Capture theory, a subset of regulatory failure theory, is the theory most trotted out by critics of regulatory agencies. If regulatory failure theory argues that a variety of factors consistently impede the agency's ability to fulfill its public interest mandate, then capture theory maintains that the agency cannot serve the public because it "systematically favors the private interests."[59] According to Horwitz, Theodore Lowi in *The End of Liberalism* blames the agency mechanism itself—the channeling of legislative and adjudicatory power to a new body—and the "vague and conflicting legislative mandate" for creating the conditions of capture.[60] Regulatory agencies are thus seen as inherently corrupt, forming "centers of private power within the state."[61]

Conspiracy theory goes even further than the regulatory failure or capture theories, arguing that regulatory agencies never exist to serve the public interest. Their good intentions are not hijacked so that corporate power can burrow into the government—the agencies are created specifically to aid industries.[62] By examining the "key participation of businessmen" in the origins of various regulatory agencies, conspiracy theorists are able to discount the public benefits of any regulatory action. At the opposite end of conspiracy theory is organizational behavior theory, an approach that traces outcomes to the specific limitations and practices of the agency.[63] Finally, capitalist state theory adopts a Marxist approach that sees the government as automatically aligned with the maintenance and success of capitalism.[64]

Captured or Not?

Although Horwitz criticizes capture theory for robbing the state of its own "essential functions" and "internal prerogatives," capture is the recurring motif in assessments of the FCC's job and outcomes.[65] McChesney argues that the FRC looked to the industry as an "ally" and essentially allowed that ally to steer radio regulation.[66] Johnson, himself a former FCC commissioner, describes the agency-industry relationship as one of domination by the "subgovernmental phenomenon."[67] The subgovernment is wealthy and "self-perpetuating," and in the field of television it includes a long list of participants who "cluster around" the FCC: broadcasters and their trade associations, communications lawyers, the various mouthpieces of the industry, and any government employee who has dealings with the industry.[68] In this view, the subgovernment, operating in plain sight, controls regulation and is one of the factors preventing the FCC from working for the public.

Krasnow, Longley, and Terry deny any such control or domination but admit the likelihood that the agency believes it needs to enable the industry.[69] Robinson, likewise, critiques the simplistic system implied by capture theory. How, for instance, would the FCC choose among the different industry voices vying for its attention?[70] While Robinson calls capture theory "awkward," he concedes that it can be an accurate predictor of whether the FCC will favor "industry [or] nonindustry interests."[71] Given the commission's history, it could also explain how the FCC would choose between commercial and noncommercial interests. Freeman, who advances a theory of "negotiated relationships," prefers a similarly middle-of-the-road conceptualization of regulator-industry relations. The regulator and the regulated depend on each other, and neither dominates.[72] In fact, Freeman envisions industries as potential sources of help and legitimation for the regulators.[73] She dismisses monolithic characterizations of private actors, arguing instead for their heterogeneous and not always insidious influence. Writing forty years earlier, Landis advances a somewhat similar assessment, acknowledging that regulator-industry meetings are "generally productive of intelligent ideas."[74] In his estimation, however, the industry's "machine-gun-like impact" on agencies has skewed regulators' priorities.[75]

Clarke proposes that we conceive of regulatory efficacy as existing on a continuum between regulatory capture and what he calls "ascendancy."[76] Ascendancy is the best possible outcome and requires that regulators have legal, political, and financial support. Only under these ideal conditions is the regulator able to "command the industry's attention" and truly lead.[77] When the regulatory body reaches this height, it is "seen as a positive source of guidance as well as of negative sanctions for misconduct, as being constructive and expressive of the public interest while alive to industry constraints, pressures and objectives, but not subordinated by them."[78] I argue that in the span of 1948 to 1952, the FCC was neither captured nor ascendant. Unfortunately, at times it was tangled up in a point on the continuum that resembled Clarke's definition of "truculence."[79] For Clarke, the industry's resistance to a regulator can be so stubbornly foul as to be "a general truculence, founded in a conviction that the regulator is uninformed and does not understand, whereas they, the regulated, do."[80] It is at this point on the continuum "where regulators get stuck."[81] As chapter 3 illustrated, and as this chapter will show, the NAB/NARTB tended toward truculence. The issue of program standards created an impasse of sorts for both the government and the industry. By examining the FCC's dealings with the NAB and its understanding of its own power, we can complicate the association's obstinance and antigovernment rancor. First, it is necessary to delve briefly into the working environment of the FCC—a factor in what might be considered the commission's weakness or pro-industry leanings.

"A Flea-Bitten Pup": Some Realities of the FCC's Working Conditions

Although most of the literature agrees on the odd, inconsistent, and incoherent duties of regulatory agencies, scholars tend to disagree on whether and to what extent the FCC has failed to serve the public interest, has been captured, or has ever worked for the public interest. However, the one thing everyone seems to agree on is that the FCC isn't doing *something* correctly. In his summation of the different regulatory agencies in 1960, Landis writes:

> The Federal Communications Commission presents a somewhat extraordinary spectacle. Despite considerable technical excellence on the part of its staff, the Commission has drifted, vacillated and stalled in almost every major area. It seems incapable of policy planning, of disposing within a reasonable period of time the business before it, of fashioning procedures that are effective to deal with its problems. The available evidence indicates that it, more than any other

agency, has been susceptible to ex parte presentations, and that it has been subservient, far too subservient, to the subcommittees on communications of the Congress and their members. A strong suspicion also exists that far too great an influence is exercised over the Commission by the networks.[82]

Vincent Mosco observes that criticisms leveled at the FCC encompass "every imaginable reason," and that some offer such conflicting reasons for the commission's faults that "one cannot help but think at times the critics must be referring to different organizations."[83] Streeter also notes the consistent barrage of negative attention directed at the FCC.[84] Such negativity did not go unnoticed by commissioners. In a speech in 1951, Commissioner George E. Sterling discussed improvements made in television reception and remarked, "The broadcasters got the praise, not the FCC. We seldom receive a letter of praise from the public when we do something constructive, but *oh boy*, how the bricks fly when they do not like our decisions. I hope you will remember that a pat on the back of a bureaucrat is like a kind word to a flea-bitten pup."[85]

Defenses of the FCC are difficult to find, but one broadcaster managed to draft one in a letter to Representative Eugene Cox (D-GA) in 1951. Countering the conservative Cox's assessment that the FCC was overstaffed, broadcaster Waldo W. Primm advised Cox to educate himself about the FCC's duties. He wrote, "I am not one who believes the FCC to be a perfect organization, and I am not one who has been the recipient of special favors from that body. I do, however, believe the FCC has been unduly criticized many times over for actions which were deemed to be of the best interests as the FCC saw it."[86]

Noll, Peck, and McGowan explain that the FCC's behavior is reactive; wading beyond the criticisms, we must look to the difficulties presented by the commission's "environment" in order to understand its behavior.[87] For example, the commission's industry orientation, or its alleged lack of concern for the public interest, could depend on who has the loudest voice. In other words, in disputes between the industry and the public interest, the public interest often has no one to speak for it.[88] This is one reason why complaints and feedback from viewers and organizations, though not representative of the entire population, are important contributions to the legal structure of broadcasting. Ostensibly, the commission should be the one speaking for the public interest, but the burden of industry pressure cannot be discounted. Another environmental factor is the FCC's independence. Noll, Peck, and McGowan argue that the commission, lacking "external review and judgment" unless specifically provoked by an invitation or appeal, is protected from "visibility and change."[89] The result is a sort of administrative inbreeding, a sealed-off policy lineage that constantly reproduces itself. But regulatory bodies are dynamic, as we can see from the dramatic change that occurred from James Lawrence Fly's FCC, which investigated the networks and published the anti-monopolist *Report on Chain Broadcasting* in 1941, to Mark Fowler's FCC, which was aggressively deregulatory and favored a marketplace approach to broadcast regulation in the 1980s. Commissions can also vary from one presidential administration to another as appointments expire, so the idea that the FCC is shielded and therefore static is debatable.

Two other environmental issues named by Noll, Peck, and McGowan are budget and workload.[90] Both of these issues appeared in FCC speeches in 1951, a crucial time for the television industry. The speeches indicate that the FCC was on a public relations mission, taking detours in speeches to address its budget crisis. Commissioner Sterling blamed delays in AM license applications on a budget cut and warned that a similar fate would befall TV applications after the lifting of the license freeze.[91] "We are so poor," he half-joked, "we could not even employ Einstein if he showed up seeking an engineering apprentice." Presenting the same case to the NARTB at its 1952 convention, Commissioner Paul A. Walker likened the industry-regulator relationship to that of the characters in *The Prince and the Pauper*.[92] The FCC-as-pauper was on track to face the onslaught of TV applications "with hopelessly inadequate funds and staff." Commissioner E. M. Webster even delivered a speech focusing entirely on the FCC's internal issues, echoing earlier remarks about the inevitable delay in TV application approvals due to a lean budget and the attendant inability to hire new employees.[93] Assuming the FCC operates under the best possible circumstances ignores its relatively marginalized position in the federal government.

Noll, Peck, and McGowan also point to the circumstances of employment as factors affecting the FCC's behavior. A number of scholars have raised the issue of appointments, job turnover, and job jumping, and all of these accounts are negative. The job itself is considered to be a popular one, adding "market value" to a commissioner once he or she leaves the post.[94] FCC appointees are understood to have few, if any, relevant skills.[95] The process of appointments is steeped in "patronage politics" and results in appointees who tend not to stand out in any particular way.[96] Candidates for the job have lacked strong feelings about industry or media in general and have avoided making enemies.[97] In a study of the makeup of the FRC and FCC from 1927 to 1961, Lawrence Lichty sketches the background of the "typical" commissioner.[98] In general terms, the usual appointee had a legal or governmental background.[99] Only two of the forty-four commissioners from 1927 to 1961 were educators.[100] Twenty-four commissioners counted broadcasting experience on their résumés (one had actually worked for the NAB).[101] Perhaps obviously but no less significantly, broadcasting talent was never represented on either commission.[102] Appointees' shared legal backgrounds consequently shaped commissioners' solutions to broadcast-related problems. Such backgrounds increased the probability that regulatory decisions would be settled in "legal and administrative rather than in social or economic terms."[103]

Former commissioner Nicholas Johnson refers to the FCC as a "graduate school for the regulatory subgovernment."[104] One reason for the FCC's high job turnover is the attractiveness of commissioners to the industry.[105] Job jumping, the practice of leaving a regulatory agency to work in the regulated industry, speaks volumes about who is right for the job in the first place. Furthermore, the probability of job jumping is presumed to weigh on the decision-making of commissioners.[106] Between 1927 and 1961, seven commissioners (or 25 percent of those who were not in office in 1961 and did not die or retire) moved on to a career in broadcasting.[107] Three of those men rose to the vice presidencies of networks—two at CBS and one at NBC.[108]

The environment of the FCC therefore educates us about the politics of getting the job, navigating the job, and eventually leaving the job. Understanding the environment further complicates the notion of capture, or even failure. By moving us away from questions about "what the agency ought to be doing," this context allows us to analyze, in Mosco's words, "what the Commission *is* doing and *how* it is going about doing it."[109] Social relations should not be ignored in such an analysis. The topic of the Television Code presents an opportunity to examine institutional and social relations in a moment of stress, specifically the overlapping occasions of the license freeze and the postwar growth of commercial television. By the time of the license freeze, the relationship between the FCC and the NAB had matured and settled into routines that did not bode well for discussions about television standards.

The NAB and the FCC: Professional and Social Interdependence

One of the factors binding the FCC to the industry is information. The two-way flow of reports and data from one body to the other is an example of what Freeman calls a "give and take" between public and private actors.[110] A 1949 NAB memo elaborates on this give and take, grudgingly revealing the interdependence of the FCC and the NAB. In this memo to NAB president Justin Miller, Ken Baker, head of the NAB's Research Department, described three ways in which the department had cooperated with the FCC.[111] First, the Advisory Council on Federal Reports consisted of a radio committee that assisted in editing FCC forms for broadcasters. Second, the FCC's Broadcast License Division kept the NAB current on license information (although Baker claimed that the published reports were not always consistent with the data in the actual license files). And third, the NAB relied on the FCC's Economic Division for data from individual stations' annual financial reports. According to Baker, the FCC would supply the NAB with its raw data, which the NAB's Research Department would process and distribute to NAB membership prior to its release by the FCC. The NAB would also give this processed data back to the FCC in exchange for the raw numbers. Baker expressed "embarrassment" at not being able to access financial data without the government's help. Although Baker and his department objected to the FCC asking stations about their money, he knew that stations' unwillingness to share that information directly with the NAB made their dependence on the FCC necessary if distasteful.

Just this one example of interdependence shows the importance of stepping back and learning more about how the NAB positioned itself in relation to the FCC. Long before the fight over the censorship of television erupted, Justin Miller set out to determine the best way to handle the regulator. A 1945 exchange of memos with his director of radio reveals that Miller wanted the NAB to think and behave more strategically. In one memo, he wrote that the NAB should be as supportive as possible of the FCC so as to accumulate enough critical capital to spend when the association felt that the commission was "misbehaving itself."[112] Asserting that the real problem with the FCC was the tendency of its legal department to involve itself in engineering affairs, Miller suggested that the government lawyers should be "left to [their] knitting."[113] The way in which Miller couched his sentiments demonstrated a level of condescension (or truculence) that scratches the surface of a much stronger criticism of the limits of the FCC's expertise.[114] While evincing an underlying disdain for the regulators, these 1945 memos represented a seemingly unified NAB strategy geared toward picking the right battles.

In practice, the NAB as a series of departments with diverse responsibilities was not unified in the least. In 1949, Miller asked the heads of all NAB departments to report back to him on how they had cooperated with the FCC and how they could improve upon that cooperation.[115] The responses varied in tone and scope. The Engineering Department memo was positive and even effusive in its report.[116] The head of the Research Department was more subdued, explaining that philosophical differences prevented their relationship from being anything other than just satisfactory.[117] Public relations and publications leaders noted very little cooperation. Although the FCC distributed copies of its docket to FCC field offices, it was up to the NAB to distribute that sort of information—as well as decisions, speeches, and important let-ters—to broadcasters.[118] The Advertising Department was negative and even sarcastic.[119] Referring to the FCC's Blue Book as the "Lemon Book," Maurice Mitchell, the head of advertising, characterized the FCC's statements about commercial broadcasting as "clumsy" and misleading. The Advertising De-partment did not even "recognize the legitimacy" of the FCC's involvement in commercial radio. The subject of censorship united the responses from the Program Department and the Legal Department. Harold Fair, head of the Program Department, wrote that "the FCC and programming are poles apart" in most cases.[120] Fair actually recommended *new* legislation to erect a distinct border between the FCC and the rights of station owners to operate without government interference. The NAB's Legal Department likewise wrote of the need to rein in a rogue FCC and educate its staff about censorship and the US Constitution.[121]

These accounts of professional relationships are incomplete without a con-sideration of the interpersonal work that took place outside of the office. Social relationships seemed to be the forte of A. D. Willard, the executive vice presi-dent of the NAB (and soon-to-be head of the NAB's new TV Department). Willard wrote to Miller about his numerous informal encounters with the former FCC chairman and commissioners, which included monthly lunches and regular outings to the golf course.[122] Yet he advocated for an "arm's length" professional relationship. He also noted the regular meals that he, Miller, and former FCC chair Charles Denny would share. Denny, of course, left the chairmanship of the FCC to work for NBC in 1947 under suspicious circum-stances.[123] Notably, Willard wrote that his attempts to set up monthly lunches with new FCC chair Wayne Coy were met with lukewarm responses. Coy would not prove to be the great friend of the NAB that Denny had been. Re-gardless of his previously full social calendar, Willard had warned of compro-mising the NAB's "principles" by getting too close to the FCC; the association had to be watchful for any abuses of power.

The exchange of memos initiated by Miller, which took place just a few months after the House select committee publicly assailed the FCC for its Blue Book, leaves us with a picture of philosophical fractures and social awkward-ness in the midst of necessary interdependence. The data- and technology-driven factions of the NAB had few complaints about the FCC because they conceived of their fields as less subjective than the others, but the area heads focused on content and legalities had a territorial and prickly reaction to the thought of cooperation with the regulatory body. Miller acknowledged as much in a letter regarding a technology that broadcasters opposed. He wrote that the FCC "would be much better employed" if it occupied itself with resolving technological issues, "instead of attempting to make philosophical, psychological, political and other changes" in programming.[124] Convinced of a foundational disconnect with the FCC and its interpretations of the Communications Act of 1934, the NAB moved into the television age feeling burned by the FCC's attempts to regulate radio.

A Public Feud and a Hidden Tension

Justin Miller had a real problem with the FCC and most regulatory agencies as a matter of principle. At the "All-Media Conference on Freedom of Expression" in 1948, he reasoned that the FCC, unchecked by judicial review, essentially could leverage the same power as a constitutional amendment.[125] Chapter 3 elucidated Miller's attitudes about government power, but his correspondence with individual commissioners also bears examination. One especially brutal exchange of letters between Justin Miller and Commissioner Robert F. Jones over color television manufacturing is notable for its visibility and for its revelation of unrest within the NAB. Unfolding in 1950—in the middle of the license freeze—the private argument rapidly became public at a tense time in the television industry.

The argument began with Miller's comments to Jones over one paragraph in a speech Jones had delivered two days earlier.[126] In his speech, Jones accused the television industry of using the FCC to delay color TV, an action that the commissioner suggested "might well lead to" an enforcement of antitrust laws. In response, Miller wrote a letter in which he sarcastically accused Jones of occupying a "happy medium" between the dictatorial practices of Joseph Stalin and the "repressive action" indicated by his invocation of antitrust.[127] Befuddled and offended by Miller's accusations of government overreach, Jones wrote back to Miller and conveyed his surprise that the NAB favored the industry's attempts to "freeze television in black and white."[128] Miller's letter confirmed for Jones "the accusations that have been made by small broadcasters [. . .] that the NAB speaks for the power segments of the industry." Not content to keep this exchange between the two men private, Jones made the letters public—a move that prompted a press release from the NAB in which Miller claimed that Jones misunderstood his initial letter.[129]

Justin Miller and others at the NAB had a history of hyperbolically painting the FCC as a dictatorship. The NAB's opinions about the FCC were clear, but less clear were broadcasters' attitudes about the NAB. The only evidence we have seen of broadcasters' discontent with the NAB has been in the fluctuation of membership numbers (see chapter 2). Those numbers indicate instability but fail to explain it. Jones's revelation that small broadcasters felt unrepresented by the NAB paints the FCC's job in a new light. A hazard awaited the commission because, as Krasnow, Longley, and Terry state, "if a regulatory commission is content to respond to dominant interests, it may lose its meaning, whereas if it defies major forces in its environment, it may risk its existence."[130] With the NAB ostensibly speaking for all broadcasters but visibly supporting larger interests like the networks, the FCC found itself in the position of negotiating the tensions between the public and the industry, between Congress and the industry, and between the powerful and less powerful *within* the industry. Intra-industry tension created more problems for the FCC, since it regulated license holders and not the networks but had to contend with the NAB, which represented both. Much to the consternation of the NAB, television programs distributed by the networks to affiliated stations became a concern for the FCC because of the responsibilities of the license-holders. The FCC's management of the growing turmoil over television programming in the face of suspicion and ire from the trade association adds another problematic dimension to the Television Code.

The FCC from 1948 to 1952: Taking on TV

A 1976 government report on FCC and Federal Trade Commission (FTC) appointments crafts a surprisingly captivating narrative about the creation of the commission with which this book is primarily concerned.[131] From 1948 to 1952, the composition of the FCC remained constant. Wayne Coy served as chair with Paul A. Walker, Rosel H. Hyde, Edward M. Webster, Robert F. Jones, George E. Sterling, and Frieda B. Hennock laboring alongside him. Coy, a Democrat from Indiana, was Truman's "strong man" and part of the president's overall strategy to install a politically adept, "energetic" leader aligned with the administration.[132]

A "cooperative" FCC rounded out Coy's chairmanship and contributed to a "stable" commission during the tumult over television's rollout and channel allocations.[133] The report's authors noted that the FCC "might have lapsed into a slumber" without television, and they maintained that the new medium factored heavily into Truman's strategy for the commission.[134] Lawyers dominated the seats, but all had some connection to or philosophical investment in the field of communication. Chairman Coy was a lawyer and a former radio station manager.[135] Commissioner Hennock, a Democrat from New York, was also a lawyer and during her tenure at the FCC campaigned aggressively for educational television.[136] Her confirmation was a surprise, given the Republican Senate's reluctance to install a Democrat in a long-term appointment in 1948, an election year.[137] Commissioner Walker, a lawyer from Kentucky and a Democratic appointee of President Franklin D. Roosevelt, had served on the commission since its creation in 1934.[138] Both Commissioner Sterling, a Maine Republican, and Commissioner Webster, a Washington, DC, independent, had military backgrounds, and both served as engineers for the FCC prior to their appointments in 1947.[139] Their appointments were also "deft political" maneuvers by Truman, as his administration had to face a Republican Senate not eager to work with a Democratic president who may have been on his way out of office.[140] Amid some controversy about his extreme right-wing beliefs, Robert F. Jones, a Republican congressman from Ohio, was confirmed in 1947.[141] Commissioner Hyde was a Republican lawyer from Idaho and entered government service in 1924, traveling from the Civil Services Commission to the Office of Public Buildings and Parks and finally to the FRC through its evolution into the FCC.[142] Hyde was originally confirmed in 1946 to fill a vacancy, and Truman reappointed him in 1952 because his knowledge of television allocations was virtually unparalleled.[143]

This FCC found itself confronting the challenge of television programming under fractious, post–Blue Book circumstances. This did not dampen the spirit of jocularity that sometimes made its way into FCC speeches. To an NAB audience in 1950, Chairman Coy, whom Justin Miller had saddled with the speech title "The American Broadcaster's Responsibility to His Government," quoted one of Miller's speeches about censorship and responded to Miller's rhetoric jovially:

> I am happy to report to you that Judge Miller has been as good as his word and he has not made any attempt to censor my speech. It comes to you uncensored, un-blue-pencilled, unedited, and probably ungrammatical. For my part, I am grateful to the Judge for his stand against censorship which permits me to speak freely here this afternoon. The Judge and I see eye to eye on that issue.[144]

Still, Coy advised the NAB membership in attendance to be cognizant of community needs, to facilitate "discussion of local issues," and to showcase "local talent"—all hallmarks of the dreaded Blue Book. Encouraging "new approaches" and "new formats," Coy reminded the audience that money wasn't everything and that "good taste and decorum" were paramount for those privileged to hold a license.

The topic of censorship appeared in other speeches as well. For his part, Commissioner Sterling pushed for self-regulation and "self-disciplining" and lambasted the presence of censorship "of any kind."[145] At the same time that he distanced his job as a commissioner from the job of controlling content (and invoked Section 326), he warned broadcasters that "justified wrath at their excesses may unfortunately lead to pressures for an unjustified censorship." Sterling identified a distinct line between acceptable and excessive, but maintained that crossing the line did not merit a legal response. The constitutionality of the FCC's efforts was no small concern. Since, as Krasnow, Longley, and Terry note, the courts had designated the FCC a "traffic cop," the question remained if that cop should manage not just the flow of cars on the road but the contents of those cars as well.[146]

The FCC and the First Amendment

Henry Goldberg and Michael Couzens connect the FCC's evaluation of "program service" to public utility regulation, which required that those granted the power to provide a utility had to operate at a "desired level of service."[147] In broadcasting, the "desired level of service" may have coincided with an abstract definition of quality based on the public interest, but the FCC did not create a consistent standard. Nevertheless, the mechanism by which the FCC attempted to maintain certain standards was the broadcast license, which, as we have seen, has stimulated arguments about whether the commission regularly steps outside of constitutional bounds. Many of these arguments fix on the licensing power itself and elide the effect of that power on the behavior of the licensees.[148] For Robinson, the FCC's "indirect influence" on programming is more dangerous than its direct actions.[149] While avoiding judgment of this influence, Rosenberg concedes that the FCC's own standards are "rarely [. . .] spelled out" and instead take the form of press releases, complaint hearings, or nonbinding opinions in hearing decisions.[150] Boylan argues, too, that the FCC's use of "speeches, informal publications, policy pronouncements, and casual dicta" allow the commission to get away with a level of control that flirts with unconstitutionality.[151] Carlton Fleming observes that since the FCC avoided creating standards, broadcasters quickly grew accustomed to scrutinizing commissioners' "out-of-school remarks [. . .] and [. . .] FCC publications."[152] As one might imagine, the NAB was incensed with this "raised eyebrow" approach to regulation. In his article, former NAB board chairman John Dille Jr. bemoans the scare tactics employed by the FCC. In one such example, the FCC "sends letters to stations, then publishes the fact in the trade press. The station manager in a small town reads of the action and alters his practice so that he won't be called on the carpet."[153]

These indirect means of establishing standards exist outside of judicial review because, as Fleming notes, they are not "agency action."[154] The flow of information between public and private actors that Freeman stresses factors into indirect influence as well, since the exchanges amount to "official, semi-official and unofficial advisory opinions" moving across regulatory and industry boundaries.[155] Furthermore, the commission's "in terrorem" control—the threat of periodic review in the form of license renewals—amounts to another form of regulation without the explicit crafting of policy.[156]

FCC speeches, one type of informal regulation, disclose the commission's real apprehension about the quality of television content. In speeches to educators, Commissioner Hennock frequently characterized television as a problem and prodded educators to avoid repeating the mistakes they made with radio.[157] Worried that television would mimic the "patterns and formulas" of radio, Hennock warned that "unwise and unintelligent programming" could lead to the "degradation of our national tastes and standards."[158] Chairman Coy took a similar tack, blaming commercialism for stunting creativity and advocating an "integral" position for "non-commercial interests."[159] A speech Coy gave at the University of Oklahoma in March 1950 set off alarms in the trade press. In that speech, he remarked, "The question of just how bad poor taste can get before it verges over into downright obscenity or indecency may be settled one of these days if the present drift in that direction is not checked."[160] These remarks echoed his warning one month earlier that failure to tone down television violence might lead to "serious" consequences.[161] According to the trade magazine *Broadcasting*, Coy denied that his Oklahoma speech carried the threat of FCC-initiated censorship and further questioned the viability of government or industry codes (though he "hinted" that a new Blue Book was "not an impossibility").[162] An *NAB Member Service* article, published less than a week after Coy's speech, attempted to contain the commissioner's negativity. The article stated that Coy "described offenders as few" and "pointed to a great mass of wholesome entertainment on air."[163]

The legal literature elaborates on the implications of criticisms like Coy's. Robinson points out the irony that the FCC's criticism of television programs ignored the ways in which the commission's own policies promoted homogenous, "safe" programming.[164] Cox blames bland programming on the "timid and uninventive" industry rather than on FCC policies, "since such policies really don't exist."[165] Going further, Cox argues that although the FCC should not establish "minimum standards of quality," it should insist that "serious entertainment" be a required programming category.[166] However, the existence of categories, which the FCC relied upon to determine whether certain types of programs were on the air in sufficient numbers, amounted to prohibition. Broadcasters needed to prove that their stations met minimum percentages of programming categories—a ritual that was completely divorced from the quality of the programming in each category but that nevertheless conceded that certain categories were preferred and others were unacceptable.[167] In applying for license renewals in the early 1950s, stations had to identify how many of their programs were live, kinescopes, commercial, sustaining, network, and locally produced.[168] The FCC's preference for sustaining, local, and live programs was implicit in these categories. Requiring broadcasters to classify their programs in this way was a remnant of the controversial Blue Book. But the FCC had learned its lesson, Robinson argues, by holding on to its principles without "stating its standards in terms of absolute demands."[169] In Loevinger's assessment, a more specific policy of category preferences—one that targeted crime programs, for instance—would cross a legal line.[170] Loevinger states that, because only a certain number of programs can make it to air, a policy of category preferences that places limits on less favorable program categories would automatically mean that some programs in that category would not air. Even a less extreme, unwritten policy of category preferences might prompt one to ask, as Goldberg and Couzens do, if the broadcaster's schedule actually reflects the community's or the commission's desires.[171]

Evidence suggests that the FCC's interest in guiding both television content and the television industry's attempts at self-regulation began to emerge even before the license freeze had ended. Early in 1951, the FCC announced that it would hold a conference to address the young medium's content troubles.[172] As a result, more rumors of a Blue Book for television began circulating. By the middle of 1951, Blue Book language was trickling into FCC speeches. The "minority tastes" that the Blue Book wanted broadcasters to tap into reappeared in a speech by Chairman Coy to the NARTB Television Program Standards Committee in June 1951.[173] After denouncing legislation proposed by Congressman Thomas Lane (D-MA) that would place the FCC in the role of censor, Chairman Coy chastised broadcasters for omitting religious and educational programs from their schedules. This particular speech is an excellent example of Coy's willingness to threaten broadcasters indirectly. Two lines from this speech should have stood out to commercial broadcasters: "The desirability of re-examining program practices by television broadcasting is brought into sharp focus by other recent developments," and "Proponents of subscription television [. . .] are certainly entitled to a careful hearing for the Commission could not afford to overlook any possibility of developing any system which may give new benefits to the public." Coy connected the dots between the specter of a decentered commercial model and the FCC's power to approve and facilitate subscription television by leading those dots through programming standards. The chairman added that subscription television could find a home with the "minority tastes" that sponsors had abandoned.

The FCC fully recognized its capacity to push the industry to adopt stronger self-regulation. One vivid example of this materialized when Coy relayed the results of a study of one week's worth of television programming.[174] Using programming "criteria" developed for radio, the FCC found that in one week in 1950, television broadcasts, on average, consisted of 72 percent entertainment programming, 3 percent educational programming, 3 percent discussion-oriented programming, 0.9 percent religious programming, and 0.2 percent agricultural programming. The takeaways from this study, Coy stated, were that viewers' religious and educational needs were going unfulfilled and that broadcasters had no idea whom they served. "It is clear that some steps must be taken by the television broadcasters to discover what the needs of their communities really are," he said. Coy's emphasis on specific categories of programming and on station-community interaction telegraphed where the FCC located the public interest and how broadcasters could modify their behavior to serve it.

In 1951, the FCC began to urge everyone to agitate for quality broadcasts. Commissioner Hennock placed the blame for low-quality programming squarely on the shoulders of the public, educators, and broadcasters.[175] She justified the absence of the FCC from that list—"The Commission cannot tell a station what to program or what not to telecast"—and confusingly dismissed "superimposition from above" while recognizing no power distinction between networks, broadcasters, and viewers. Commissioner Webster also tried to rally the public. Since the FCC could not concern itself with program content, he encouraged viewers to form community radio and television councils to create a feedback mechanism for local stations.[176] But Webster asked audience members to be mindful of the business of broadcasting. Because stations needed to make money, they could not indulge everyone's requests for sustaining programs or local programs in prime-time slots. Commissioner Walker distanced the FCC from broadcast content and, as the others did, called upon "responsible citizens and organizations" to lead the fight for better programs.[177] And because Walker delivered his speech to a religious organization, he framed television in terms of good versus evil and noted that only broadcasters with a strong "character and sense of public responsibility" could contribute to the good that television could do. *Television Digest* even reprinted excerpts of Walker's speech because of increasingly "dubious programming," mounting viewer complaints, and the commissioner's uncharacteristically "sharp" response to these circumstances.[178] The trade magazine interpreted Walker's comments as an endorsement of the nascent Television Code.[179] Justin Miller had an identical interpretation and wrote to Walker to share the good news that the NARTB board had approved a draft of the Television Code at its most recent meeting.[180]

Chairman Coy remained doubtful that a typical industry code would fix the problem, and he was especially concerned that a minimally reworked Motion Picture Production Code would lack the specificity necessary to confront a technology that reached into the home.[181] In June 1951, Coy suggested to NBC vice president Frank M. Russell that a committee to monitor stations and send copies of the meeting minutes to all broadcasters would be preferable to an NARTB code.[182] Coy encouraged network- and station-created standards, but he was pessimistic that a trade association code would be enforceable. Robert Swezey, who headed the NARTB's Television Board of Directors, agreed that a committee would be more effective than a code filled with "a lot of generalities virtually impossible to administer and enforce."[183] Although Swezey said that the Television Program Standards Committee should heed Coy's suggestion, minutes of that committee's first meeting show that Coy's idea was not raised.[184]

The NARTB moved ahead with its code plans, and in late October 1951, after repeating the ritual of admonishing television programs (which had "slid deeper into public distaste") and explaining the FCC's rights to "consider [. . .] program policy" but not to censor, Commissioner Sterling patted the NARTB on the back for taking "a strong step forward on their own behalf in self-discipline."[185] Like Coy, Sterling managed to incorporate a threat into his comments on the soon-to-be-adopted Television Code: "TV broadcast licensees cannot continue to turn a deaf ear to the complaints of the public; neither can the Commission. The broadcasters have the opportunity to act first." The implication that the FCC would act next if the broadcasters failed to fix the problem confirmed the Code's usefulness, but it also chipped away at the FCC's insistence that it adhered strictly to Section 326.

Commissioner Walker, too, praised the NARTB's adoption of the Code and particularly the Code's acknowledgment of the special protection owed to children and their educational needs.[186] And while he repeated the call for audience councils, he slapped the public's hand for being too puerile. After referencing the quantity of complaint letters the FCC received, Walker cautioned,

> It is easy to point the finger of criticism and scorn at broadcasters. While of course they must accept their share of the responsibility, it is unfair and oversimplifies the problem to put all the blame on them for inferior program service. The fact is that networks and broadcasting stations have been looking for a formula which would hold the attention of the great number of people and to a large extent the average level of radio and television programs reflects our immature wants and interests quite as much as it fosters them.

Such a scolding tone contradicts remarks that Walker had made two months earlier in *Broadcasting*, when he stated that the Code was a "constructive effort [. . .] that is more responsive to the needs and wishes of the people."[187]

Unlike the NAB speeches discussed in chapter 3, the commissioners' speeches failed to commit to much. They smacked of timidity, as though the FCC was still smarting from the Blue Book backlash. Their threats lacked teeth, and their public interest prescriptions lacked vision. However, although they appeared vague and benign, we know the power embedded in their informality. One can imagine the broadcasters scouring the text for hints about changing standards for license renewals. The indirect influence exerted through these speeches helped to catalyze the industry. The NARTB, responding to the possibility of another Blue Book, scrambled to present as unshakable "free" television's dominance in the face of alternative distribution methods (in other words, subscription television). An unsettled atmosphere swirled as the license freeze showed signs of thaw. The only options for TV content seemed to be an industry code or a piece of legislation. All loose ends needed to be tied up if television was to take its place alongside—and perhaps even above—radio.

Conclusion

Earlier in this chapter, I mentioned the constant criticism that targeted the FCC. In 1953, Paul A. Walker, assuming once again the chairmanship of the FCC, had this to say on the subject:

In our efforts to develop a broadcasting industry that operates in the public interest we on the Commission have been subjected to constant criticism and even abuse. [. . .] So you will sympathize with me when I tell you that for nineteen years I have been on the receiving end of a barrage from those who view with alarm, and those who wring their hands in despair. [. . .] On the one hand we are criticized by the listeners. They complain to us vigorously and continually about everything from low necklines and off-color jokes to crime programs and commentators. They object when their favorite "private-eye" program is interrupted by a national political convention. With equal fervor they protest if their favorite candidate isn't on the air. The constant demand is that the Commission do something about programming. And what happens every time that the Commission does try to do anything about improving programming? Does that make everybody happy? Of course not. We are given a sound verbal beating by the industry and particularly by the trade press. We are belabored with charges of censorship, socialism, and government domination. If we had actually ruined the broadcasting industry as often as we have been accused of doing so, radio and television really would be in a bad way.[188]

Given the entirety of the commission's remarks presented in this chapter (even Walker's exasperated protests), we should ask if the FCC had a direct connection to the Television Code. In other words, did the NARTB's Television Code masquerade as a private initiative when, in fact, it was an example of state action—government regulation by proxy? Daniel Brenner advances a two-step strategy by which the FCC could have forced the Code into existence. In this hypothetical exercise, the FCC would first "prod the industry to adopt practices that, but for self-regulatory preemption, would likely have been imposed by" the Commission.[189] It would then "applaud the industry's somewhat unspontaneous *fait accompli*."[190] Here we can tie the indirect influence discussed earlier to the crafting of the Code and not just to the worries of individual broadcasters. Brenner argues that the government played such a sizable role in the genesis of the Code that the self-regulatory document "does not lie entirely beyond the reach of the First Amendment."[191] Although it is difficult to prove state action, Brenner does argue that the NAB in general transcends its private role and functions as a "quasi-public" supervisor.[192] In Freeman's theory of interdependence, state action becomes even more difficult to prove. But Freeman also suggests a deeper level of interaction between public and private actors. When, for example, the NARTB used the threat of government intervention to plant the seed of a Code, the NARTB was trading on the FCC's authority. "Simply by doing nothing," Freeman writes, "the agency may help bolster a private trade association's authority to regulate member firms."[193] Netzhammer finds that the Television Code was not, in fact, an example of state action, but the literature in general is divided.[194] In the next chapter, we will see a clear example of government intervention and the strong reaction it elicited from the NARTB.

CHAPTER 6

Senator William Benton Challenges the Commercial Television Paradigm

NAB/NARTB speeches and publications are littered with references to government coercion and overreach. The FCC was the frequent target of the association's vitriol, but lawmakers, too, stoked anxiety in the industry and triggered abusive backlash. In 1951, H.R. 3482, which proposed to amend Section 326 of the Communications Act of 1934—the section prohibiting the FCC from censoring—faced considerable pushback from both the NAB and the FCC. The amendment sought to grant the FCC the power to bar from television "programs of any language, sound, sign, image, picture, or other matter or thing which is obscene, lewd, lascivious, or otherwise offensive to public decency."[1] In addition to authoring this bill, Representative Thomas Lane (D-MA) proposed on the House floor in February 1951 that Congress allow the FCC to constitute a censorship board "to scrutinize every telecast in advance, and to cut out all words and actions that arouse the passions, or that hold up any individual, race, creed, group, or belief to mockery and derision."[2] Referring to television as a "juvenile delinquent," Lane essentially blamed the medium's liveness for rendering parents incapable of controlling the family's entertainment.[3] In lieu of destroying television, which Lane admitted would not solve anything, the federal government needed to sanitize it.

The year 1951 proved to be a banner year for outrage over this new medium's liberties, even though (or perhaps because) a Supreme Court action had dealt a blow to would-be censors of television. In 1949, when Pennsylvania, Ohio, Kansas, Maryland, New York, and Virginia all had censorship boards for motion pictures, television's First Amendment protections were tested. The Pennsylvania State Board of Censors decreed at the beginning of 1949 that any film intended for broadcast must be reviewed and approved prior to its airing.[4] Citing a conflict with congressional authority, "an unreasonable burden upon interstate commerce," and the violation of the First Amendment, five television license holders sued the Board of Censors. In *Allen B. DuMont Laboratories, Inc., et al. v. Carroll, et al.*, the judge found for the plaintiffs based on the strength of their first two arguments but withheld comment on the third. As the *Journal of the Federal Communications Bar Association* points out, Judge Kirkpatrick was content to avoid the censorship issue, although he did accept the "conclusion of law" that "television, like newspapers and radio, is included in the press whose freedom is guaranteed by the First and Fourteenth Amendments to the Constitution of the United States."[5] In the final analysis, the court found that the Pennsylvania censors' regulation, if instituted, would be inconsistent with the existing policy passed by Congress and executed by the FCC. Furthermore, the process of submitting all film to the Board of Censors would have strained the tightly scheduled and tightly financed process of television production and programming.[6] That burden would have extended to advertisers, who might have turned away from television as the regulation increased the cost of doing business. That the commercial rather than the artistic aspect of television helped rescue it from state censorship is not inconsequential. Motion pictures would have to wait until 1952 to escape the type of censorship that TV successfully avoided.[7] The regulation of television was the FCC's territory, and state censorship boards could not trespass upon the commission's congressional mandate. However, what seemed like a win for the television industry turned out to be an invitation to congressmen to foreground the FCC's powerlessness and correct it.

In May, FCC chairman Coy warned the NARTB Television Program Standards Committee, the committee established to draft the Code, that "no less than ten bills [. . .] dealing with television programming and practices" were circulating.[8] One of these bills was introduced by Senators William Benton (D-CT), Lester Hunt (D-WY), John Bricker (R-OH), and Leverett Saltonstall (R-MA). William Benton was the face of the bill known simply as the Benton Bill in industry circles, and he is at the center of this chapter. A former advertising executive–turned–educational radio and TV advocate, Benton was a complicated character. His mission to win channel allocations for educational broadcasters and his desire to see a viable subscription television system posed an existential challenge to the industry and created some complications for most commissioners at the FCC. Benton championed causes that antagonized the existing commercial structure, and the visibility of his causes expands the way in which this book has discussed content thus far. For Benton, content regulation was not to be confined to the evaluation of decency. Legislators and regulators needed to rethink the prevailing definition of TV and entertain the possibility that the medium could accommodate strong educational content, as well as niche varieties of content delivered commercial-free via subscription. As it turned out, the most visible element of Benton's multipronged approach to improving broadcasting was the creation of what the industry considered to be a national censorship board for radio and TV. Strategically, the Benton Bill was a misstep, a gift to the NARTB that allowed the association to obscure Benton's progressive initiatives by singling out the specter of censorship. The Television Code has long been seen as a response to threats from legislators like Benton and Lane, but as this chapter will make clear, it was equally—if not more importantly—about preserving and naturalizing television's commercial framework.

The Complicated Figure of William Benton

William Benton served in the US Senate from 1949 to 1952, initially as an appointment to fill a vacant seat. He had launched his career in advertising in the late 1920s; cofounding his own advertising agency with Chester Bowles in 1929, he remained in that partnership until 1936.[9] In her article chronicling Benton's transition from adman to radio reformer, Cynthia Meyers writes that in 1933 Benton began voicing concern about "programming quality," arguing that catering to advertisers and not listeners degraded a show.[10] He even supported ejecting advertisers from program-building altogether. He proposed that networks should control the content and "sell interstitial minutes to advertisers."[11] According to Meyers, Benton worried that the hypercommercialization of radio "alienated audiences."[12] Advertisers simply had no investment in the public interest. Benton believed that networks were the key to reinforcing the public interest clause because as organizations that "worked with both advertisers and local stations, [they] were better positioned to oversee a process of editorial responsibility, a process that could ensure that programming was designed at least in part for the audience's benefit."[13] Benton's sights were set on upending the model of sponsor-controlled programs, but as he looked to the networks for solutions, he also considered alternative programming and distribution options.

Meyers writes that Benton, a corporate liberal, wanted to shield "the public from [the free market's] worst excesses."[14] He pursued two objectives that he felt might secure this protection. First, Benton promoted the role of education in "building a better world."[15] As a vice president at the University of Chicago and as subsequent owner and operator of the *Encyclopedia Britannica*, Benton's conspicuous career shift demonstrated his commitment to championing the public interest in its various forms. The bridge that spanned his old life in advertising and his new one in education was radio. Educational broadcasts could reach people who did not have access to institutions of higher learning. To that end, he developed educational radio programs at the University of Chicago, but to hear him tell the story in 1951, he "failed, miserably, in the goal [he] had set for [himself]—and for radio."[16]

Benton's second objective involved an alternative business model for radio. He placed even more distance between his new career and his old one when he pushed for a subscription form of radio that would sidestep a commercial presence altogether.[17] This proposal was unpopular with commercial broadcasters, but Benton stood his ground, rightly pointing out that advertising had become the de facto way of doing radio even though nothing in radio's technological makeup required commercial support.[18] The subscription model had been introduced as early as 1924 with the creation of the Radio Music Fund Committee, which solicited donations to fund music programs. The American Society of Composers and Publishers (ASCAP) believed that this model would yield quality programming, but this discourse clashed with the industry-supported conviction that "the supposed freeness of content" was "part of broadcasting's essence."[19] Sewell reminds us that this principle drew a straight line from "broadcasting's indirect economics" to "democracy and uplift."[20] We can, in turn, draw a straight line from industry discourse of the 1920s to the language and deeds mobilized to stop William Benton when he set his sights on television.

"Television with a Conscience": Benton and the Potential of TV

In his article "Television with a Conscience," which appeared in the August 25, 1951, issue of the *Saturday Review of Literature*, Benton warned that the impending end to the FCC's license freeze would confine television to a paradigm dictated by advertisers. Decrying "program stereotypes" and "lowest-common-denominator" options, Benton proposed three ways to stop TV from proceeding down "the road to trivialization."[21] First, networks and stations needed to air more educational programs. Second, a workable subscription TV service needed to compete with the advertiser-supported model. And third, the government should grant more schools licenses to operate TV stations. Benton bemoaned the overabundance of pure entertainment and crime formulas, arguing that if the medium continued along this path, "without guidance from Congress or from organized public opinion, [it] will never remotely do the great and urgent educational and public service job required by the times."[22]

Benton believed that the only way to fix the education problem was to fix the commercial problem. He asked, "Why not a system in which the viewer orders and pays for what he gets—without any advertising whatsoever?"[23] Framing subscription TV as both a home-based movie theater (he referred to a "box office [. . .] at the receiving end of television") and an educational institution (he referred to the service bill as a "tuition fee"), he envisioned subscription TV not just as a public service but as a special venue for niche audiences, whom he referred to as "minorities."[24] While commercial television catered to a mass audience, subscription TV could offer "class" in the form of opera, plays, and other high-culture forms.[25]

Benton encouraged his readers to stay abreast of the FCC's allocation of TV channels for educational broadcasters. Of the 2,000 stations the FCC would authorize, Benton stated that 209 had been "tentatively reserved for 'non-commercial, educational' use."[26] He praised the FCC for the reservations, but he warned that the "powerful" NARTB not only opposed the reservations but predicted that the channels would be "wasted through non-use or partial use."[27] Benton had already seen evidence of this sort of opposition to noncommercial and educational radio.[28]

The NAB had fought against educational television initiatives like Benton's at FCC hearings on educational TV back in January 1951. At that time, Kenneth Baker, the NAB's director of research, used license and grant figures to conclude that radio had been lost on educators.[29] "Once granted a license to broadcast," Baker stated, "an educational institution is more likely to let it lapse or lose it by default or disuse than it is to develop the grant into an effective contact with the people it serves." Baker argued that educators' failures to use radio and even motion pictures adequately would only lead to more of the same in television. He concluded by conceding the possibility of successful educational stations, but asserted that "educators as a group have not evidenced the willingness nor the competence in using radio that would justify the reservation to them of any part of the broadcast spectrum." What the NAB conveniently ignored was that broadcasters ushered in commercial radio well after amateurs, the government, and schools had established a presence on the air.[30] Educational TV was represented in those hearings by, among others, the Joint Committee on Educational Television (JCET). While the NAB pointed to what it perceived to be educational broadcasters' failures, the JCET, too, used the "record of commercial radio" to argue against the ability of the commercial broadcasting model to fulfill the public interest mandate.[31] Aware of the NAB's strategic arguments from these hearings, Benton was prepared for a similar fight.

Benton knew that the struggle against the commercial domination of TV would be alleviated if he could find a solution to the funding problem. So he recommended alternative funding mechanisms, one being the commercial sponsorship of a half-day's worth of programming in order to support the other half's educational shows. Another method was, again, subscription TV, whereby, for example, a university could sell access to its football games on a pay TV channel and, with that money, offer educational programming for free year-round.[32] Firmly attached to his goals for educational TV and an alternative distribution strategy that would support it, Benton pursued a policy that would provoke the wrath of the NARTB and test the FCC's independence.

Benton's Legislative Maneuvers

In "Television with a Conscience," Benton suggested a fourth solution to the TV problem that would "buttress" his three initial objectives: the passage of S. 1579, which included the creation of a National Citizens Advisory Board for Radio and Television. But Benton had to tiptoe toward S. 1579. He and Senator Lester Hunt made an impression first with S. Res. 127, a necessary starting point for his vision for TV. Introduced on April 13, 1951, S. Res. 127 was a proposal to study educational television and was concerned with TV's overall impact on all aspects of US life.[33] Noting the overwhelming response to the televised Kefauver Crime Commission hearings in March, Benton called for Congress to "take a sharp look" at television and to "set general policies" for both the FCC and the industry to follow. The scope of Benton and Hunt's proposed study would be expansive, encompassing the amount of airtime allotted to public service programming, the FCC's criteria for channel allocations, commercial sponsors' influence over programming, and government financing of educational television.[34] By conducting this type of study, Congress, Benton believed, would hold "a key, a hope—a chance to get at the basic problem of our times, which is the understanding and education of our fellow citizens."[35] Benton was careful in his remarks before the Senate to underscore his expertise, emphasizing his involvement in the "two programs which did the most to open the eyes and pocketbooks of advertisers to the enormous selling power of broadcasting": *Amos 'n' Andy* and the *Maxwell House Showboat*. By folding into his narrative the seemingly contradictory roles of advertising executive and educational broadcasting advocate, Benton skillfully painted himself not as an anticommercial convert but as a realist, a believer in the public interest with the vocabulary of a commercial insider.

He also positioned himself shrewdly as a radio man and not a television man. He stated that he was "staggered by the promise—some call it a threat—of television." He believed that television was too potent to be ignored by Congress and "relegated [. . .] to the custody" of an administrative agency like the FCC. Benton was far too dismissive of the FCC's power, especially considering that the commission had issued a notice of unassigned television channels just one month prior. Benton predicted that the rush for these channels would result in "the real freeze," the establishment of "a mold and a pattern" of the type of broadcasting already dominating television. He acknowledged sponsors' strong influence on programming, which he believed yielded little more than good entertainment. Moreover, he worried that educational programming would find no home during peak viewing hours. "I am not objecting to the Bing Crosbys and Bob Hopes of television of the next two decades," Benton remarked. "I merely suggest that it is in the public interest that their programs be supplemented—and not merely in the early morning hours or late at night after most of America is in bed—by the kind of programs to which all of us want our children exposed, even if we do not spend too many hours listening to them ourselves."

Benton also challenged the FCC's plan to allocate only 10 percent of the unassigned channels to educational broadcasters. As with the conflict over noncommercial radio frequency allocations that McChesney's work explicates, educators sought substantially more than 10 percent of available TV channels. Seeking an allotment of 25 percent, educators exposed a rift between the government's outlook and theirs. Benton saw the allocation disparity as a disconnect made only less comprehensible by the popularity of the televised Kefauver hearings—programming that could have been classified as educational.

Two issues here merit elaboration: the availability of educational programming and its scheduling. As we saw in chapter 4, scheduling was of tremendous consequence for sponsors. The reliance of sponsors on profitable scheduling actually prevented NBC from abiding by its own programming standards. That Benton demanded not only the availability of educational programming but also its optimal scheduling speaks to his industry knowledge as well as to his lack of qualms about control over the flow and content of TV.

Well versed in the expensive and labor-intensive process of production, Benton was careful to address the criticism—made most recently by the NAB before the FCC—that educators had failed to exploit radio fully. He called this criticism unfair: "The educators have not been seriously asked to try to do a job with radio broadcasting. On the contrary there has been tremendous effort and much money expended to keep them out of it and to keep them quiet. Very few of our leading educators have taken any active role in the study or development of radio broadcasting. They have lacked not only the facilities but they have lacked the money." This direct jab at the NAB and the networks allowed Benton to highlight the financial chasm that separated commercial and educational broadcasters. It also provided the perfect segue to the topic of subscription television ("box office in the home television") as an alternative platform and funding mechanism. Pushing back against the accusation that a subscription model was antithetical to the "so-called American system of broadcasting," Benton argued that such a model was the very embodiment of "free enterprise," since it would sell itself directly instead of "giving [itself] away," as in the commercial model.

When he answered his critics and stated his case, Benton introduced his resolution, which targeted the interconnectedness of television's power, radio's procommercial past, and entrenched corporate interests. Speed was essential, Benton pointed out, because policies (and biases) carried over from radio could "harden into [. . .] an unfortunate pattern" that would reveal itself in a channel allocation plan that favored commercial broadcasters. The *Congressional Record* reflects Benton's submission of two articles on the benefits of subscription television, a fitting end to the senator's case for an alternative to a model that had used the public's airwaves but, in his estimation, had failed to operate in the public interest.

Reactions to S. Res. 127

A win quickly followed Benton's April 13 remarks. After securing three educational television stations for his home state of Connecticut—two more than the original allotment—Benton's office issued a press release in which he stated his case to a wider audience than the one that had heard him just a few days earlier.[36] He praised TV as a "marvelous new instrument of human communication" and agreed that its use for entertainment was a positive development, but he was captivated more by its potential to "lift the level of citizenship in this country to the highest point ever achieved by any nation." Benton was eager to spread the word to the general public, emphasizing that subscription television would ease the burden on taxpayers, but he also sought feedback from commercial broadcasters, as indicated by an exchange between himself and CBS president Frank Stanton shortly after his remarks on the Senate floor. Clarifying to Stanton that his primary concern was "programming in the public interest" and not "transient public tastes," Benton differentiated his interests from those of the television industry while avoiding antagonistic rhetoric.[37] A few days later, he prodded Stanton for his "informal reactions" to his speech and resolution, revealing that Senator Ernest McFarland (D-AZ), chairman of the Senate Committee on Interstate and Foreign Commerce Subcommittee to Study and Investigate Radio, Telegraph, and Telephone Communications (SCIFCS), had "promised" there would be hearings on the resolution.[38]

Chairman Coy reacted to the resolution even before Benton introduced it, but the senator did not take note. Coy stated outright that he "would have to oppose it in any hearings."[39] The proposed allocation delay was out of the question, as the FCC's plans "were based on three solid years of work, including much bitter and acrimonious discussion." At no time, Coy pointed out, did Congress intervene, even though they were kept informed. Amending any part of the allocation plan would mean disrupting all of it, a move that had international consequences. "We would be holding up the Canadians and Mexicans as well," Coy stated, "because they have been waiting for our allocations before they made their own." The allocation of 209 channels to educators was, in Coy's opinion, "adequate," and he was quick to remind Benton that the Kefauver hearings the senator anchored his resolution to "were all on *commercial* stations."[40] Even Coy, a progressive chairman, could not envision the elaborate engineering scheme of channel allocation accommodating the demands of a content-driven resolution. Once again, regulators characterized TV service solely as a technical hurdle, so they likewise framed their solutions as commonsense matters of course rather than as value-laden decisions that excluded other voices.

Accordingly, and without even having read Benton's resolution, Commissioner Walker foresaw "great opposition to this in the industry." Not missing a beat, the NARTB's Television Board of Directors drafted a resolution in opposition to S. Res. 127 at its Chicago convention on April 17, and soon thereafter Justin Miller wrote to Benton to set up a meeting to talk "in detail" about the resolution.[41] In response to news that Senator Edwin C. Johnson (D-CO), chairman of the Senate Committee on Interstate and Foreign Commerce, wanted to hold hearings on S. Res. 127, the NARTB invited its members to think about Benton's resolution as part of a "long-term campaign" against commercial TV.[42] In its bulletin to member stations, the NARTB supported its theory of an anticommercial campaign by linking S. Res. 127 to the Ford Foundation's recent $90,000 grant to the JCET. The bulletin did include a caveat that "the two developments are not necessarily related directly," but the NARTB attempted to provoke a degree of envy by drawing members' attention to the fact that $90,000 was "approximately equal to the total television budget of NARTB at this moment for all activities."[43] Despite appeals to conspiracy theories and outright jealousy, the bulletin quoted the association's official response to Benton's resolution (crafted in the form of a resolution at the NARTB's Chicago conference), which argued that the investigation was essentially redundant, that the Communications Act of 1934 clearly established the criteria for FCC licensing, and, most importantly, that the investigation would postpone the lifting of the television license freeze.[44]

Here we see a significant economic motive for upbraiding Benton: the freeze had stalled the expansion of television service. Any delay would compound the losses that the television industry had been incurring since the beginning of the freeze in 1948.

Clearly incensed, Benton brought the NARTB Television Board of Directors' resolution to the Senate's attention on May 15. In his speech on the Senate floor, he lamented the power that commercial broadcasters wielded in Washington and regarded with incredulity their fierce opposition to the few reservations the FCC had allocated to educational stations (209 in total).[45] Irate that the TV board's resolution had targeted S. Res. 127 "for the singular honor of condemnation," Benton seized the opportunity to elaborate on the role of television in education.[46] Benton and Hunt then conducted an exchange on the Senate floor pointedly rehashing the early monopolization of radio licenses by private corporations and providing vital context for the opportunity presented by the allocation of television channels. In a reversal of his previously low opinion of the FCC, Benton even called for a "strengthening [of] the hand" of the commission in reserving and maintaining channels for educational use.[47] He skillfully voiced his support of subscription television (a "new commercial development," he called it) just before he requested that the FCC apply the phrase "public interest, convenience, and necessity" specifically to television.[48]

On the topic of licensing and content, Senator Kenneth S. Wherry (R-NE) argued that speech made possible by government license was distinct from speech in a newspaper. Benton agreed, in a way establishing a baseline for the encroachment of the government upon ideas disseminated by radio and television.[49] But instead of traveling down that particular path, the path to censorship, both senators leaned on the scarcity argument, advocating for greater diversity of access to those precious licenses, which restricted the breadth of speech, they argued, in ways that a newspaper or a street corner soapbox did not. When one party received a license to broadcast, another party did not. It was up to the government, then, to make licenses available to those without the financial and political heft of the commercial broadcasters.

While Benton navigated the Senate floor, parsing the implications and importance of his and Hunt's resolution, John Howe, Senator Benton's legislative assistant, solicited advice about S. Res. 127 from several well-placed experts. The resulting counsel was not hopeful. Howe ascertained that Senator McFarland would "grant [Benton] a courtesy hearing" and then "forget the whole thing."[50] Furthermore, Benton was warned that "the chance that any new legislation would ultimately emerge" from the resolution (if passed), would be "almost zero." Howe also discovered that Benton's base of support was compromised. Educational organizations like the National Association of Educational Broadcasters and the JCET could not rally behind Benton because, in accepting money from the Ford Foundation, they had to swear before the Bureau of Internal Revenue that they would not attempt to "influence legislation." Commissioner Walker, too, conveyed pessimism to Howe because of the FCC's inability "to withstand pressure even with new legislation behind it." To illustrate his point he cited one case in which a congressman with a vested interest in a license application that was denied retaliated against the FCC by alleging that it had granted a license to communists. The FCC wound up losing over half a million dollars from its appropriation. The shared interests of elected officials and industry stakeholders meant that, in Howe's words, "on the Hill industry will be prepared to make a real fight." Benton, in turn, was prepared to provoke the industry.

The May 31 SCIFCS hearings on S. Res. 127 yielded more information about the circumstances surrounding the resolution than trade publications and other available documents divulged. In his opening remarks, Benton relayed that one of the criticisms leveled at him by the industry centered on his timing.[51] Although the senator had been active in educational pursuits since his retirement from advertising in 1936, his decision to throw television channel allocations into the legislative fray in 1951 came a bit too late for his critics. His response was that he was unaware that "great organizations" would involve themselves in the public interest potential of TV to the extent that they had. One can assume that he was referring to the Ford Foundation, an organization that could not publicly align itself with Benton. In fact, just over a week before this hearing, Benton wrote to Paul Hoffman of the Ford Foundation to discuss the urgency of reserving channels for educational purposes, securing good time slots for educational programs ("between six and ten p.m."), and developing a subscription TV service.[52] In this letter, Benton envisioned a supportive relationship between the Ford Foundation and the FCC. He wrote, "Except for you, I know of no other group representing the public interest apart from the FCC itself. The FCC is beset with hundreds of problems and in my judgment could hardly fail to appreciate guidance and public support." In his letter, Benton also championed the expansion of FCC powers and possible amendments to the Communications Act of 1934.

Benton's testimony on May 31 built on the sympathy he evinced in his letter to Hoffman. His treatment of the FCC in the hearing was respectful and even forgiving. Questioning Benton's aims in holding up channel allocations for up to one year, Senator McFarland asked if he felt that the FCC was not suited to handle the technical side of broadcasting, specifically the channel allocations. Benton was quick to praise the expert body but acknowledged astutely that industry pressures on the commission—which ultimately affected the public—required the attention of Congress. Only a minority had bent the FCC's ear regarding the reservation of channels for educational use. The majority informing and persuading the FCC had been representatives of the commercial side of broadcasting, and Benton explicitly mentioned the NARTB's opposition to S. Res. 127 and the paltry 10 percent allocation. Benton actually spoke highly of the FCC's ability to secure that 10 percent allocation. Given the NARTB's negative reaction to the plan, Benton confessed that the FCC showed "nerve and courage."[53]

As the hearing wore on, Benton did not adhere solely to the issue of allocations for educational broadcasters. He also announced that he was ready to introduce a new, more specific resolution as well as a piece of legislation. He spoke of his new bill, which would amend the Communications Act of 1934, as one devoted primarily to creating "higher standards for television."[54] He then introduced the idea of an "annual Blue Book," which the bill would facilitate.[55] This type of annual review would be geared toward educational and public service programming, but when pressed by Senator Johnson for his definition of "educational program," Benton hesitated to paint too broad or too narrow a picture. That reluctance to settle on a definition would return to haunt him. Nevertheless, Benton went on to explain that one result of the passage of his bill would be a clearer understanding of "public service" programming, a "much used and abused phrase."[56] He did not profess to have all the answers, but he did formulate a plan for somehow elevating television.

Subscription television was one part of that plan. When asked why a subscription service would result in better television, Benton replied that radio and television quality had stagnated because of commercial sponsors. So, too, had diversity of programming. Networks and sponsors ignored niche audiences— ranchers who wanted to learn about wool or lawyers who wanted a review of Supreme Court decisions—because programming to them would not be profitable. For Benton, the public interest, education, and commercial broadcasting were all connected in a negative way. His blueprint for saving television from the patterns that had marginalized public service programming in radio began with S. Res. 127 but crystallized with S. J. Res. 76 and S. 1579.

The Benton Bill

Benton and his cosponsors, Senators Hunt, Bricker, and Saltonstall, submitted S. 1579 on May 31, 1951. The composition of the sponsors was politically savvy. Farthest to the left of the political spectrum was Benton, with Hunt occupying a more centrist Democratic position.[57] From the conservative and liberal wings of the Republican Party came Bricker and Saltonstall, respectively.[58] Bricker typically looked to the free market and not to the government for solutions to the country's problems, but as we see here, he aligned himself with educational and noncommercial broadcasting interests.[59] He was especially important to the cause because, as Perlman notes, his conservative reputation helped to deflect charges of antidemocratic behavior.[60]

Despite the fact that the bill had four sponsors, S. 1579 became known simply as the Benton Bill. The crux of the Benton Bill, and the reason for the NAB's apoplectic response, was the bill's creation of a National Citizens Advisory Board for Radio and Television. Made up of eleven "outstanding private citizens, drawn from the fields of education and communications and from among leaders in the civic, cultural, and religious life of the nation"—with not one having any financial interest in radio or television—the Citizens Advisory Board (CAB) would "review how well radio and television are serving the public interest" and issue an annual report to Congress and the FCC.[61] The CAB's most pressing task would be to "suggest how radio and TV could serve the public interest better."[62] Benton made it clear that the CAB would not supplant the FCC, but he did assume that the CAB would influence legislation and offer the FCC guidance on license renewals.[63] He also emphasized the CAB's advisory role, which he felt would protect it from accusations of censorship. "Yet," he wrote, "its potency should not be underestimated."[64]

The industry and the trade press clung to three major facets of Benton's plans: the "annual Blue Book," the prolongation of the license freeze, and the threat of censorship. An article in *Broadcasting* described some broadcasters' fears of being "subject to constant surveillance and whim of a semigovernmental body" in the form of the CAB.[65] And the article noted that although Benton had received an overwhelmingly positive response from educators, "some 'unsolicited' individuals" had written to Benton opposing a "super-censorship board."[66] We should not gloss over his letters of praise from citizen reformers, as they disclose those aspects of the Benton Bill that were most important to his supporters. Mary Graves, who "would not have [a television set] as a gift on a silver tray," was attracted to Benton's plan for a subscription service.[67] Peter Perry simply wanted to see action taken, and he criticized the FCC for not "rid[ing] herd on the broadcasters."[68] William Elliot was pleased that programs would finally be monitored, and Morris Goldstone, drawn to television's educational potential, proposed that a state-run channel could defeat communism.[69] Beth K. Spencer admired Benton for fighting "for the purpose of safeguarding the rights of the citizenry against triviality and commercialism," while William L. Gordon Jr. admired him for fighting to "safeguard the quality of T.V. programs."[70] Benton also attracted the attention of organizations such as the North Wisconsin Lake Superior Education Association, the Wisconsin Association for Better Radio and Television, and the Woman's Christian Temperance Union.[71] A righteous crusader to some, Benton found much-needed support among the types of citizens who had sought television reform since the medium's earliest days.

He could not count on the same support from the FCC, which lacked a unified opinion of the bill. Commissioner Hennock, a vocal proponent of television's educational uses, favored a congressional investigation of the criteria for channel allocations, noting that an FCC-initiated investigation was "unlikely."[72] Chairman Coy was not pleased with the bill's call for a year-long delay in channel allocations; in a letter to Senator Johnson, Coy wrote that such a delay "would [. . .] harm the public."[73] Although Coy and Benton diverged on the issue of the license freeze, they converged on the necessity of programming standards. However, Coy wanted to create the impression that the FCC had the programming issues in hand, since the commission had announced public hearings on television programs to be held at an unspecified date.[74] Regardless of the FCC's ambivalence toward the bill, Benton had drawn the commission into what the industry incorrectly perceived to be a power grab. Industry representatives took them both to task.

A striking letter from Eugene Carr, the NARTB's director of radio, to Senator Bricker initiated a string of ugly accusations and exemplified the association's anger toward government involvement in the industry's affairs.[75] In his letter, Carr expressed relief that Bricker did not agree entirely with Benton's plan. He went on to outline the "real danger" of the bill—the "supercensorship board for all radio and television." He accused the FCC of "trying for years to grab complete control of programming in radio" and saw the CAB as a way for the commission to accomplish that with congressional approval. In a crafty turn, Carr posited that Benton's advocacy of subscription broadcasting, a serial failure in Carr's assessment, was a consequence of his personal investment in Muzak, a nonbroadcast music service.

After accusing Benton of corruption, Carr reiterated one detail from his conversation with Senator Bricker the previous week—that the senator's "real interest" in the Benton Bill was the allocation of a television channel to The Ohio State University. Carr assured Bricker that, based on his research, the FCC had indeed allocated a channel to that location. "There can be no question in mind," he wrote, "but what this channel is earmarked for Ohio State, so it would appear that your primary interest in this respect is covered." The channel was reserved in the ultra-high frequency, but that did not matter, Carr argued, since most television stations would soon be operating in UHF. Carr neglected to mention that most television sets were incapable of receiving

UHF signals, and he likewise did not relay the tremendous opposition to UHF that RCA had mounted.[76] Carr concluded the letter by assuring Bricker that his needs had been met, a circumstance he presumably hoped might curb the momentum of the Benton Bill. Any momentum was questionable, however, since Senator Johnson had assured Justin Miller in mid-June that the Interstate and Foreign Commerce Committee was "not apt to take any very speedy action" on the bill and joint resolution.[77] Miller had every indication that the Benton Bill would be "just as 'fresh' in August" when Miller returned from vacation as it was in June.[78]

Freezing out the CAB

Carr decided that a so-called super-censorship board was the most threatening aspect of the Benton Bill, but from an economic perspective the most pressing problem was actually the provision that would prohibit the FCC from allocating television channels for another six months to one year. Benton defended the allocation ban, explaining that educators were not as savvy and organized as entrenched radio and television interests; they required more time to acquaint themselves with the realities of television broadcasting.[79] The SCIFCS wanted more information, however, and held hearings on July 18 to look into the particulars of the license freeze.[80] At those hearings, Chairman Coy told the committee that the NARTB had petitioned the FCC on July 8 to halt freeze-related hearings in order to hasten the freeze's end.[81] The FCC complied, and barring any objections to the plan to dispense with oral hearings and address only written comments (of which there were 1,200), Coy predicted that the freeze could be over by the end of September 1951. Notably, on July 25, Coy filed a letter with the committee relaying the news that his prediction of a quick thaw was "somewhat silly."[82] The revised end-date for the freeze extended well into November. Coy's optimistic prediction aside, the takeaway from this development was the NARTB's attempt to end the freeze prematurely with the FCC's blessing, before all oral hearings could be processed, thus eliminating one of Benton's mechanisms to insert educational broadcasters into the new channel allocation scheme. In other words, the NARTB attacked on multiple fronts, tapping into individuals' fears of government overreach (as discussed in chapter 3), massaging relationships with key elected officials, and attempting to work the regulatory bureaucracy to its advantage.

The SCIFCS license freeze hearing was lengthy and meandered through a number of related topics, but it concluded with the subject of allocations for educational use. While contemplating the costs and benefits of having universities make little to no use of their allocations versus having commercial broadcasters devote some of their airtime to educational programming, Senator Johnson raised a question about the FCC's programming authority. The question highlighted one central, persistent debate about the FCC's power and helped Benton position his bill as both viable and constitutional. Senator Johnson asked Coy if the FCC had the power to mandate that a percentage of a licensee's airtime be devoted to educational programs, and Coy replied that he did not believe that was so.[83] He believed that dictating programming would violate Sec. 326 of the Communications Act. Johnson replied that if the FCC could distinguish between educational and commercial stations, thus dictating the type of programming of one station versus another, then it could technically "reserve part of a channel" for educational programming.[84] Coy was not convinced of Johnson's argument, differentiating between the "nature" of a class of programming—safety, military, commercial—and the "character" of the programming.[85]

In a memo to Coy, Max Goldman, the acting general counsel of the FCC, wrote that the commission could find a basis in the Communications Act for dictating percentages of types of programs to licensees, but doing so would reverse the pattern of licensee responsibility long ingrained in broadcasting.[86] Had the FCC assumed those powers from the beginning, the story would have unfolded differently. Another issue to consider, the lawyers proposed, was the specificity of community needs, to which the licensee was supposed to be attentive. A percentage system dictated by a federal agency could not possibly account for the needs of each community. Any alteration to the FCC's existing practice would require new legislation. A competing memo to Coy from Benedict Cottone, FCC general counsel, countered that the FCC had every right to implement a percentage requirement.[87] Simply because the FCC had not wanted to engage in such a practice in the past did not mean it lacked the authority to do so, Cottone asserted. "It is my view," he wrote, "that censorship is no more involved in this matter than the setting up of a class of stations specifically for broadcasting only educational programs." Casting aside the "desirability" of the FCC establishing programming quotas, Cottone stated that no legal issue obstructed the path. Nevertheless, in the hearing Coy characterized the obstacles to implementing a percentage plan as "formidable," citing as the first major problem the task of firmly establishing the criteria for labeling a program "educational."[88]

At this point Benton interjected, arguing that a percentage plan would not solve the problem.[89] Any applicant for a television license would acquiesce to such a mandate simply because the drive to have the license superseded any limitations placed on its use. Over time, however, the profit motive would supersede the educational requirements, and the broadcaster would relegate public service programs to the margins of the programming day. Benton's reasoning led him to the conclusion that only legislation would correct the trends already established by the commercial radio system. His presence at the freeze hearing—notable because he was not a member of the committee holding the hearing—and his insertion of his legislative prerogatives into the record saw him matching the NARTB tactic for tactic.

Expedient Revisions

Eventually, Benton became convinced that the language and demands of his bill were too antagonistic. In late July, he and the bill's cosponsors revised the bill and resolution, removing the section on the license freeze and clarifying that the CAB was not a censorship board and thus would not assume a role similar to that of the FCC. In submitting the revisions to FCC commissioners Jones, Sterling, and Webster for approval, Benton mentioned Coy's "general approval" and expected hearings to commence in August.[90] Benton sent nearly identical letters to the heads of ABC, NBC, and CBS and explained to all that the revisions aimed "to make it crystal clear that the function of the proposed National Citizens Advisory Board for Radio and Television is strictly advisory."[91] Although the industry worried that the CAB would be a quasi-governmental body, creating redundancies with the FCC, John Howe stated in a meeting with NARTB leadership in July that Benton's setup was purely practical. A completely private CAB was not possible, owing to a lack of funding sources. The federal government alone would be able to "provide not only the machinery, but the funds."[92] At this meeting, Harold E. Fellows, by this time the NARTB president, assured Howe that Benton would be met only with the NARTB's "firm opposition."[93]

In August 1951, Senator Benton and his cosponsors introduced revised drafts of S. 1579 and S. J. Res. 76, both of which replaced Benton's original resolution, S. Res. 127. If adopted, S. J. Res. 76 would recommend that TV stations continue to be licensed annually, that the FCC pursue the development of subscription television, and that the CAB take shape.[94] Benton explained in the hearings, which took place on September 5, that upon realizing the enormous pressure that the industry was under—pressure from private citizens as well as from businesses—to bring television service to the wide swaths of the country that had limited or no service at all, he and his cosponsors had opted to backtrack on their freeze extension provision.[95] As noted earlier, the principal revision to the Benton Bill pertained to the CAB. Benton stated that the bill's changes were "designed to emphasize (a) that the proposed board is to be advisory only; (b) that its functions are in no way to duplicate or overlap the functions of the FCC, and that the functions of the advisory board must in no way derogate from the exclusive authority of the FCC to grant, withhold, or revoke licenses."[96] These changes corrected what Benton described as the "misunderstandings" that were prompted, in part, because he "unfortunately referred to the Blue Book" when he first testified.[97] As discussed in chapter 1, the Blue Book quickly became the equivalent of an obscenity in industry circles, and Benton conceded that he "made the worst possible mistake [. . .] by identifying something positive and constructive that in the minds of the industry is a bad symbol."[98] Unable to make the point strongly enough, Benton further explained that he had "no thought that either the Federal Communications Commission exercised censorship over the content or form of television or broadcasting, or that the proposed advisory board have any power, in any sense, as a censor. Its power will be educational and formational, rather than any statutory authority in the field of censorship."[99] Additionally, the CAB would not conduct individual station reviews. It would simply "review trends or point to examples of good or bad programs"—a charge similar, I should note, to that which the Blue Book attempted to execute.[100]

The new bill further amended its call to study particular broadcasting problems by calling instead for a "study of programming trends" as they related to the problems of "community needs," educational and cultural programming, "the financing of broadcast operations," and "the groups which exercise effective control of programming."[101] The CAB, meanwhile, would still report to Congress and the FCC annually and advise on potential legislation or any related matter.[102] Benton further testified that the new bill was "generally acceptable to Chairman Coy."[103] In fact, Coy was "generally" in favor of a committee in the vein of the CAB.[104] According to *Broadcasting*, Coy had remarked that "it would be a good thing to know whether the people are getting their money's worth."[105] He subscribed to Benton's premise, but he insisted on couching it in terms of consumer, rather than citizen, welfare.

The issue of censorship hovered over the hearing, however, and Benton raised it again while waiting for a witness to arrive. Entering into the record a critical editorial from *Broadcasting*, Benton protested against the accusation that the CAB would operate illegally as a censor but sympathized with the industry's trepidations, though he noted later that he had no evidence of any particular instance that would have triggered fear on the part of the industry.[106] The CAB, Benton argued, would not have one member "confirmed who believed in censorship or who took an appointment on the board for the purpose of acting as a censor."[107] Senator McFarland asked the obvious question: "Well, how much good would the board accomplish without in some way censoring these programs?"[108] Benton's response rested in his belief in the power of opinion leaders to effect change in an industry completely resistant to it. In his mind, an annual report would lead to editorials, which would lead to congressional debates, which would then trickle down to the practices of individual stations.[109] McFarland probed further and asked if the CAB's recommendations should inform the FCC's own decision-making. Hedging a bit, Benton admitted that he would want the FCC to adhere to the CAB's reports on the occasion of a "flagrant failure on the part of a licensee to live up to the promises [. . .] made" when licensed.

At this point, Benton turned on the FCC slightly, pointing out that since the creation of the Federal Radio Commission, only two or three licenses had ever been revoked.[110] A strong CAB might give the "kind of men normally appointed" to the FCC the "courage" to take assertive action against violations.[111] Benton hypothesized that the commissioners would allow the CAB's recommendations to inform their decisions.[112] The mere threat of cooperation would then "influence" the station's behavior.[113] Nothing in Benton's reasoning, then, deviated from the established FCC practice of regulating via patterns of judgment and informal remarks rather than the formulation of actual rules. It is worth noting, too, that Benton was keenly aware of the commission's "confidence" problem. A crisis of confidence has the potential to infect all regulators, as legal scholar Clarke makes clear.[114] Susceptible to the industry's truculence and to fickle politics, regulators need to find confidence in order to stay relevant. The idea that the CAB might bolster the FCC's self-esteem is simultaneously intriguing and dangerous. Benton's hypothesis all but bellowed that the commission, an independent agency created by Congress to oversee all of communications, was impotent.

Convinced that his fixes were sound, Benton tried to sell the revision to a wide audience, publishing "Television with a Conscience" in the *Saturday Review of Literature* and even reaching out to network heads whose interests he knew were not always synchronized with the NARTB's (as discussed in chapter 2). In a letter to Robert Kintner of ABC, Benton tried to make the case that the CAB would help the US image overseas. "The feeling of foreigners about American radio [. . .] could hardly be worse," he wrote.[115] He further reasoned that a strong CAB would "greatly enlarge" freedom of expression, and he added that the existence of such a board earlier would have made radio a freer medium. The goal of the letter, however, was to recruit Kintner to support the cause by "tak[ing] the kind of leadership that is so desperately needed in the industry."

Benton's outreach to the industry had no obvious positive impact on its trade association. Not long after Benton's testimony, the NARTB Television Board sent statements to Senators McFarland and Johnson, calling the bill

FIGURE 6.1. "Boo!" (cartoon by Sid Hix, 1951)
Though exaggerated, the response of the NARTB TV board members in this Sid
Hix cartoon is not out of tune with the panic expressed in NARTB documents.
Cartoon by Sid Hix for *Broadcasting-Telecasting* (December 3, 1951). Courtesy of
Broadcasting & Cable.

"potentially more dangerous to free expression than any legislation that has
been before Congress in the thirty-year history of broadcasting."[116] And at
every NARTB district meeting throughout the month of September, members drafted resolutions opposing the Benton Bill. Congratulating its members for the display of unity, the NARTB declared in its bulletin, "This is the
type of cooperation which builds strength and stability into an industry and
protects it from infringements."[117] The NARTB also had other tools at its
disposal, as events in September demonstrated.

Fabricating a Scandal

In the September 1 issue of *NARTB Member Service: Confidential TV Newsletter* sent to NARTB members, the lead story was dedicated to Benton in

light of the upcoming hearings on S. J. Res. 76 and S. 1579. Characterizing the CAB as "one of the most potentially-devastating legislative guns ever leveled at the radio-television industry," the piece calls for all broadcasters to unite in opposition to legislation that could lead to a "super-censorship board."[118] Siding with viewers in its denunciation of S. 1579, the editorial states, "The rights of the mass of viewers and listeners would be abrogated to the extent that a super-board would determine that certain televised and broadcast programs liked by viewers en masse were not proper for them; the board would state this conclusion annually to Congress."

The article then takes an accusatory turn, first by pointing to an underlying conspiracy. The article notes that the revised bill was drafted by former FCC chair Paul Porter and Porter's law partner Harry Plotkin, who previously worked as a lawyer for the FCC's Broadcast Division. Since Porter was FCC chair from December 1944 to February 1946, when Benton was pursuing his subscription radio plan, and Plotkin was instrumental in the drafting of the FCC's Blue Book, alarm bells sounded. This anticommercial lineage inflamed suspicion.

Next, the article attacks Benton's background and motivations. Benton got his seat in the Senate, the article reminds readers, thanks to his former advertising agency partner Chester Bowles, who, when governor of Connecticut, tapped Benton to fill a vacated Senate seat. After calling into question Benton's credentials, the article goes on to use Benton's words against him, citing passages from "Television with a Conscience" and doubting the sincerity of Benton's disapproval of radio's development. For the article's author, Benton's transformation from advertising executive to television reformer was made less believeable by the fact that Benton was "sitting on a 20-sponsor wallet instead of a pillow."

Finally, the article bluntly accuses Benton of corruption. According to the article, Paul Porter, in his role as a lawyer for Muzak (which meant he worked for Benton), was resurrecting Benton's failed subscription radio plan by filing a petition with the FCC. Because Benton was a proponent of subscription television for the purposes of improving the educational function of television, and because Benton was a major shareholder in Muzak, and finally, because S. 1579 offered a path to a subscription model, the association cried foul. The NARTB board, acting on its belief that the Benton Bill was "highly damaging to the private telecasting industry," committed all its resources to "the battle" against it.[119]

For Porter, the *NARTB Member Service* article was part of a "smear" campaign.[120] In a contemptuous letter to NARTB president Harold Fellows, Porter called the NARTB's accusations against Benton "reprehensible" and asserted that the association's failure to address criticisms leveled at the industry "may invite really unpalatable remedies." He went on to write, "Serious proposals by responsible people should not be met by slick innuendos or vilification. Nor should this vital question of public policy be obscured by unwarranted charges of 'censorship.' The industry has a duty to oppose censorship, but it will endanger its future if it cries 'wolf' recklessly and heedlessly." Porter's exasperation underscored the efficiency with which the NARTB could mobilize and trigger its propaganda mechanism.

Now at the center of a wholly constructed scandal, Benton wrote to Justin Miller (by this time chairman of the NARTB board) expressing his amazement at the "shortsightedness" of the NARTB article.[121] He wrote, "I cannot comprehend it. I know a lot about trade associations, and I know how shortsighted many of them tend to be, in their efforts to appeal to the emotions and prejudices of their members. But you are in a field [. . .] where leadership, and a long range viewpoint, is essential to the ultimate prosperity of the industry." In his reply, Miller disagreed with Benton's assessment of the NARTB's reaction.[122] He wrote, "Our Association was fulfilling its highest obligation to its membership and to the people of the country in sounding the alarm bells, with vigor, when your legislative proposals, S. J. Res 76 and S. 1579 came before the Congress." He relayed the NARTB's extreme feelings:

> As Mr. Fellows stated in his letter to Paul Porter, we are genuinely convinced that the proposals of S. J. Res 76 and S. 1579 constitute "distinct threats to the nation's free media—the newspapers, magazines and movies as well as radio and television." We regard your legislative proposals as an entering censorship wedge to be exercised through indirection and endangering the freedom of all communications.

On Benton's point about shortsightedness, Miller retorted, "We believe that our 'alert' to the broadcasters revealed [. . .] *long*-sightedness into the past and a reasonable degree of *far*sightedness into the future, concerning your proposed legislation."[123] He concluded by assuring Benton "that in the presentation of our case before the Senate committee, in opposition to your proposals, we shall proceed dispassionately, and that our presentation will not be obscured by petty controversy."

Despite Miller's assurances that the NARTB would proceed "dispassionately," the NARTB's statements about Benton's conflicts of interest were so stinging that the senator wrote to the bill's cosponsors to assure them that he did not seek to profit from his support of educational and subscription television. He wrote, "Mr. Harold Fellows [. . .] assured me over the telephone today that the NARTB has had no intention of implying that I have a personal financial stake in these issues; and that, both personally and on behalf of the NARTB, he accepts unequivocally my assurance that I have not."[124] The NARTB's troublemaking gesture did not seriously compromise Benton's mission, but it did have the power to distract and to manufacture doubt. For the association intent on keeping the commercial system intact, even the hint of reform required swift, and even underhanded, action. Here we can refer to Clarke's contention that "all regulation is politics."[125] His account of a regulated industry's willingness to retaliate against "overformal" regulation rings true in the case of the NARTB.[126] Having seen a future that would surpass the industrial and regulatory status quo, the association launched a "counterattack" by inciting its members via a malicious public relations campaign.[127]

Support for and Opposition to the Benton Bill

Shortly after the offending bulletin was sent to NARTB members, the NARTB Television Board met in Virginia to deal with, among other things, the Benton Bill. At that meeting, the Television Board "adopted unanimously a resolution opposing the adoption of these measures and authorizing the NARTB to register such opposition and take such steps as may be necessary and desirable to protect the American system of television broadcasting and freedom of expression."[128] "American" clearly meant "commercial," although the NARTB rhetoric framed the bill and resolution as "threats to free speech."[129] Board directors once more characterized the CAB as being "one step away from an actual censorship agency."[130] One line in the *Broadcasting-Telecasting* article about this meeting is worth noting. After a report on the board's concerns about subscription TV and a discussion about working with educators (ostensibly to avoid a legislative solution to the educational programming problem), the article states, "Obviously the board felt it should set the pace for a determined battle against legislation deemed highly damaging to the private telecasting industry."[131] Gone was the pretense of doing battle against legislation that encroached on freedom of expression. Even an earlier line in the article points to the facade. Describing the board's "double-barreled attack" on Benton, the article refers to the first shot as the "adoption of [an] explosive resolution condemning the Benton measures and *calling them* threats to freedom of speech."[132] The best way to win this fight was to brand the resolution and bill as threats to freedom, while strategizing about the actual threats, which clearly were educational TV and subscription TV. The association seized on the opportunity to employ prodemocracy rhetoric in the thick of the Cold War.

The NARTB's claims of industry unity aside, reactions to the Benton Bill were not necessarily uniform. *Broadcasting* predictably and consistently criticized Benton's initiatives. The publication even ran an article about an angry letter from the general manager of a station in Illinois who accused Benton and his cosponsors of imposing elitist values on the practices of "sincere Americans," a criticism not foreign to proponents of educational TV.[133] Other reactions ranged from measured to supportive. Frank Stanton at CBS was not happy with the bill, even after it was revised. He did not cry censorship, but he did question the efficacy of the CAB, given the established regulatory powers of the FCC.[134] An NARTB bulletin reported that the majority of FCC commissioners opposed the CAB, with Chairman Coy and Commissioner Walker dissenting.[135] The majority of commissioners worried about censorship and a "possibility of conflict" with the FCC's own operations.[136] The majority also expressed some degree of frustration with the bill's presumptuousness. For example, Benton wanted a one-year license period for television broadcasters, and the commissioners responded that this was already the case. Additionally, they implied that the bill's provision might interfere with the FCC's "discretion in setting license periods."[137] Furthermore, the FCC had already proceeded with subscription television trials; the bill's push for a subscription TV initiative simply reiterated the mandate in the Communications Act to pursue new uses of radio.[138] Benton brushed off the FCC's feedback, convinced that the disagreements were based simply on differing points of view.[139] He believed that he would have the opportunity to answer the FCC in subsequent hearings, but *Broadcasting* reported that the Commerce Committee had made no plans for future hearings.[140]

Although the majority of FCC commissioners refused to go along with the Benton Bill, Coy, Hennock, and Walker were important allies of the senator; so, too, was the American Civil Liberties Union (ACLU). The ACLU had issued a statement early in October 1951 supporting S. 1579 and the CAB. The statement read, "The ACLU is ever vigilant to any threat of censorship and we would be the first to register our opposition to this measure if we believed that censorship control over programs was the motive behind the bill."[141] The motive, as the organization understood it, was the creation and maintenance of educational television stations. The ACLU praised the CAB's potential contribution to diverse ideas and issues and lauded the section of S. 1579 pertaining to the role of the CAB in studying sponsor control over programming. The statement vigorously denied that the bill harbored any intention to censor programming or to supplant the regulatory authority of the FCC.

Benton wrote to Patrick Murphy Malin of the ACLU shortly thereafter and mentioned that the NARTB insisted on arguing that the CAB "represent[ed] an 'indirect threat' of censorship," in spite of the ACLU's carefully drafted statement to the contrary.[142] While the ACLU stood firmly behind the bill's service to education and the public interest, Malin did raise questions about the supposed inclusiveness of a subscription-based model of broadcasting. Benton pushed back against Malin's concerns, working to dispel in particular the idea that the cost would be prohibitive for some viewers. Malin's point was "surprising" to Benton because subscriptions were not new. He further argued that the purpose of subscription TV should be to increase the "total amount of information and education afforded to the community." Having an alternative distribution method that would exist alongside the commercial model and serve "large minority groups" could accomplish that.

Meanwhile, the Television Code

The newly constituted TV board of the NARTB met for the first time on June 4, 1951, five days after Benton and his cosponsors introduced their resolution and bill.[143] By June 11, the TV board had initiated a review of station programming logs from the previous month.[144] On June 22, the Television Program Standards Committee held a daylong session to discuss the state of TV programming, with both FCC chair Wayne Coy and Senator Edwin C. Johnson speaking.[145] At the meeting, Paul Raibourn of Paramount Television Productions warned, "I would advise that we take to self-regulation immediately. The experience in the motion picture industry has been that if political censorship gets into the picture, it's impossible to get it out."[146] Senator Johnson assured NARTB members that censorship was not "looming."[147] Johnson even mentioned Benton by name, recounting the senator's inability to define education in the SCIFCS hearings and questioning his capacity to present solutions related to education.[148] Coy, meanwhile, presented attendees with a list of steps that stations could take to ensure that they were broadcasting in the public interest. Implicitly rejecting Benton's language, Coy stated that the FCC had not established quotas for public service programs.[149] The result of that meeting was a resolution to draft standards in order to "improve the character of television programming and insure its observance of good taste."[150] That draft eventually became the Television Code, which was accepted by the NARTB membership in October 1951. The chairs of both the House and Senate Commerce Committees weighed in on the Code's adoption. Representative Robert Crosser (D-OH) was "pleased" that self-regulation had preempted government intervention, and Senator Johnson, while reserving judgment about the future success of a "voluntary plan of censorship," was also glad that the industry had adhered to "American tradition" by policing itself.[151]

Charles Denny at NBC was relieved to have appeased the politicians, but even the prospect of strong self-regulation did not quell his anxiety about some of the structural changes Benton was pursuing. Denny was not troubled by what was (supposedly) leaving the air; he was worried about what Benton wanted to put *on* the air. The addition of public service and educational programs was a lingering and "far more serious" concern.[152] He identified three groups pushing the public service agenda: churches, educators, and "members of Congress who can't afford to buy television time but who feel it is important that they have an opportunity frequently to appear on television." Denny's cynicism was not baseless. Politicians needed television because *attacking television* made for good press, and *being on television* could help with policy proposals and reelection. As much as elected officials benefited from prime airtime, so did the networks. One of Denny's main complaints about public service and educational programs was their potential to disrupt the schedule. The proposal to require commercial television stations to schedule educational programs in the evening, according to Denny, would "really mess up a good program schedule."

The schedule as a highly constructed arrangement between sponsors and networks was a crucial structural reminder of how efficiently the commercial model had latched onto and helped to create the routines of daily life. During the license freeze, when television was not profitable, the networks were under the thumb of the "sponsor time franchise," an arrangement in which an advertiser colonized a particular time slot indefinitely.[153] So vital was the schedule that when television became profitable after the license freeze, the networks began chipping away at the sponsors' control of the schedule in order to create logical and profitable program adjacencies.[154] Once the networks gained control of the schedule, they managed to use it as a bargaining chip when negotiating with, and ultimately muscling, telefilm producers that wanted good time slots for their programs.[155] We have seen how the schedule played a role in NBC's attempt to manage more mature programming, and we have also seen how sponsors wanted no part of that arrangement. The schedule was even a sensitive subject for the letter-writers of chapter 4. Benton was familiar with how the programming day was broken up into specific blocks (also known as dayparts), and he knew what a financial boon the prime-time hours were for the networks. The cheaper hours were in the daytime. Prime time was the moneymaker. For Benton to suggest that some of that time be occupied with public service programs was, at best, arrogant and, at worst, a substantial threat to the commercial logic of dayparts pioneered in radio long before television's launch. Denny was right to fixate on the schedule because it symbolized in many ways the greatest source of anxiety for the industry with regard to television content: the decentering, dismantling, or even supplanting of the commercial system.

While NBC was figuring out how to save its programming slate, discourse surrounding the Benton Bill continued to focus on censorship. An article in the *New York Times* in October 1951, the month of the Code's adoption, raised a salient point about the CAB that had been ignored by the NARTB. Supporting the NARTB's right to oppose the CAB, the article stated, "The idea of having a panel of citizens sit in judgment on a mass media is very old and almost a stock remedy among well-intentioned reforms."[156] It went on to argue that an advisory board created by legislation could no longer be merely an advisory board: "It obviously has a quasi-official status not enjoyed by a private group." The selection process, the members, and the power all came into question as well. Assessing the bill, the article concluded that "the aims are so appealing that it is easy to overlook the unsound principle upon which the measure rests."[157] The problem that the article points to—the "quasi-official status not enjoyed by a private group"—was precisely the problem with the NARTB.[158] As a self-regulatory body with such a tight relationship with the FCC, the NARTB, especially in adopting the Television Code, had been accused of acting with the sanction of the state. And if the tenets of the Television Code were supported, adopted, facilitated, or otherwise blessed by the FCC, then any enforcement of the Code's programming restrictions could have been a very real case of censorship.

The "American" aura of self-regulation seemed to shield the Code from any suspicion of impropriety. In fact, the Code was thought to be such a brave show of self-discipline that any other attempt to handle content would most certainly cross a legal line. *Sponsor*'s late 1951 analysis of the Code mentioned the threat that Benton posed to the industry (solely in terms of censorship, of course) and considered the possibility that a successful Code would end all legal actions against the industry. The article asked, "Now that the code is finished [. . .] the question may be put, sans malice, whether Benton himself, or any distinguished private citizen-critic of his nomination, would dare on their own to go further than, or as far as, this NARTB code goes? One presumes to doubt it. For them to do so [. . .] might well arouse profound Congressional suspicion."[159] For reasons explored throughout this book, the Code debuted with a strong, no-nonsense brand—one that appeared to supplant Benton's best efforts to police television.

The NARTB's swift drafting, approval, and adoption of the Television Code left Benton unable to digest what he perceived to be a shameless public relations strategy. In a letter to Wayne Coy, he wrote that the Code set "impossible" goals for television stations. Moreover, he wrote that the Code and not the CAB appeared to be the real censor.[160] The ACLU agreed, calling the Code "stifling and illegal censorship."[161] In response, Justin Miller, chairman of the board of the NARTB, called the ACLU "a bunch of sneaking hypocrites."[162] The same day Benton wrote to Wayne Coy, he sent another letter to Maurice Mitchell, correctly predicting the outcome of the entire issue: "The danger of this code is of course that the Congress will be glad to accept the protestations of the industry, as embodied in this code, as the kind of excuse politicians seek—to avoid facing up to a hot and difficult issue, an issue which promises few political rewards and many political hazards."[163] Benton's sentiments parallel analyses of communications legislation from the Radio Act of 1912 to the Communications Act of 1934. Congress did not anticipate or provide for technological advancements that would compromise existing policy. Congress was content to let the industry and engineers solve ideologically loaded problems that masqueraded as technical ones. And Congress had few qualms about delegating responsibility to a regulatory body composed of unelected officials. The argument could be made that, in the case of the Television Code, that regulatory body took the form of the NARTB working with the support of some in the FCC.

The Fate of the Benton Bill

After the NARTB drafted, revised, and passed the Television Code, it sent copies to the FCC and the Senate Interstate and Foreign Commerce Committee, which was deliberating Benton's bill.[164] Nothing happened when the 82nd Congress adjourned on October 19, 1951, though Benton did address senators the next day as the session closed. During his address, he cautioned senators to be wary of "pressures they may be subjected to from broadcasters upon return to their home states."[165] Sure enough, the November 12 NARTB bulletin called on its members to "take advantage of your national legislators being home during the Congressional recess . . . make it your personal responsibility to discuss the issue with these gentlemen, and give them the broadcaster's viewpoint."[166] For its part, the NARTB leadership sustained its antagonistic stance over the holiday recess. In a speech in December 1951, NARTB president Harold E. Fellows predictably referred to the Benton Bill as "dangerous."[167]

Even Senator Johnson did not remain silent during his break. In a *Variety* article, he criticized government encroachment on private enterprise.[168]

Television was still a new medium, so the inevitable "mistakes" the industry was making did not require government meddling. His support for the industry was clear when he stated:

> To conform to the standards of good manners, good taste, and good programming, it is not necessary to destroy the sparkle in exciting real-life drama and spirit of melodrama. It does not require necklines to be raised until the poor damsel is choked; it does not demand that all humor should be attired in Sunday clothes before it is televised. These things require an application of common sense and straight thinking. Nor will any Government edict remedy any deficiency in program quality. Yet, there are those who think so.

He then took on the CAB, accusing it of being, in essence, "a sort of snooper-duper Monday-morning-quarterback society to be superimposed upon the FCC and the radio and television industry." Johnson imagined a hypothetical downward spiral for the CAB, despite the best intentions of its creators. Capable of being overtaken by its own ambitions, an advisory board like the one Benton proposed would soon enough cast off its advisory role and perhaps even clash with the FCC, leading to confusion within the industry. Sensing the growing power of the CAB, broadcasters surely would approach the board directly with programming questions. "This I submit is federal censorship!" Johnson exclaimed. Government could not please everyone who had a problem with television, and it surely could not "provide the incentive for the development of this art." He concluded by exalting the new Television Code and its application of the "American way." Johnson's insistence on an industry solution framed Benton's fight not only within jingoistic terms but also within the most accessible terms—decency and quality. He wrote, "One cannot legislate honesty. One cannot legislate character or quality." But Benton's fight was about alternative paradigms, education, and the public interest. These were wonky matters that could not compete with the more sensational censorship of low necklines. Johnson's pro-industry stance did not bode well for the future of the Benton Bill.

Although a date for the 1952 hearings was not set, the NARTB had asked to testify.[169] Senator McFarland wrote to Eugene Thomas, chairman of the NARTB Television Board, back in September 1951, assuring him that the NARTB would be heard, but also suggesting that the association would not be able to voice its opposition until 1952 since the 82nd Congress would be adjourning in October.[170] By March 1952, the month the Code went into effect at participating NARTB members' stations across the country, Miller appeared less worried by the prospect of a Benton victory. A letter from Miller to a station manager in March 1952 assured the recipient that all was "quiet on the 'Hill' with respect to the Benton proposals."[171] Miller remarked that it was simply a matter of waiting for the session to end. And it did. Three months later, Senator Johnson had convinced Miller that movement on Benton's proposals would be slow, so Miller communicated to Fellows that no "immediate action" on the part of the NARTB would be necessary.[172] Inaction at the government level was met by swift action at the NARTB. The Code launched, and the Benton Bill died.

Benton saw his bill as a solid attempt to weaken the commercial stranglehold over television, but the NARTB viewed it as yet another annoyance to defeat on its way to naturalizing commercial television. In an NARTB publication aptly titled *Scoreboard*, the association listed a number of "challenges," from theater TV to limitations on the height of transmission towers.[173] One such challenge was Benton himself. Three columns, titled "The Challenge," "NARTB's Position," and "The Outcome," reduced the NARTB-Benton struggle to wildly simplistic categories. Three sentences further reduced the story to little more than a competition. In the "Challenge" column, the scoreboard emphasized Benton's attempts to censor; in the "NARTB's Position" column, the scoreboard briefly rehashed the association's efforts to reveal the "danger of censorship" to the public and to members of Congress; and in the "Outcome" column, the scoreboard noted the death of the bill in committee. The NARTB had dispensed with Benton via a vocal adherence to stable, attention-getting talking points and via a whisper campaign designed to discredit the senator's intentions. At this moment, as in the battle against educational television, the association had to operate publicly to defend its trade from paradigms that threatened commercial dominance. Shutting down alternatives was part of the endgame, which was to shut down criticism and talk of alternatives. The Code needed to be a show of force because it needed to end the discussion. Commercial television needed to become routine. It needed to be a non-issue. It needed to be natural.

Benton's Epilogue

In 1959, long gone from the Senate and motivated by the recent quiz show scandals, Benton wrote a letter to President Eisenhower and sent copies to a number of friends.[174] In his letter, Benton reminded the president of the proposed CAB and attributed its end to his unsuccessful senatorial campaign in 1952 rather than to the efforts of the NARTB. Benton wrote that a CAB issuing annual reports on the state of television would have created a landscape in which the quiz show scandals never would have materialized. He also wrote that the CAB would have strengthened "the hand of the FCC." The real goal of his fight in 1951, the letter implies, was to chip away at the dominance of commercial entertainment but certainly not to eliminate it. "The American people have never been exposed to the alternatives," he wrote. We can sense that even seven years after the failure of his bill and his political career, Benton was still chafing at the efficiency with which the industry foreclosed the possibility of alternatives. In the aftermath of the quiz show scandals, the president had a "remarkable opportunity [. . .] to appoint the board of review which the times demand," Benton wrote. Interestingly, a line from an earlier draft of this letter, stating that this new board should not have industry professionals on it, was modified to read, "if your advisors feel that representatives of the broadcasting industry itself are needed, I would recommend the inclusion of practicing broadcasters rather than owners or administrators."[175] Burned by broadcasters' national mouthpiece and its lobbying apparatus, Benton conceded that potential for reform might exist at the local level.

The same day Benton wrote to Eisenhower, he corresponded with journalist William Hard, pleading with him to contact the White House to make sure his letter found its way to the president.[176] Three days later, Benton received a reply from Eisenhower. In a brief response, the president thanked Benton for the letter and assured him that upon his return to the White House he would have a conversation with his staff about an advisory board.[177] Buoyed by Eisenhower's professed interest, Benton fired off a series of letters—all including his exchange with Eisenhower—to associates who would help launch the new board. Clearly inserting himself back into the fight, Benton wrote to Senator Thomas J. Dodd (D-CT) and asked if he would mention the Eisenhower exchange to Senator Saltonstall, one of the original bill's sponsors.[178] "I don't want any publicity of any kind," Benton wrote. "I'd just like to get the commission set up and the job started. [. . .] This chance offered by the President is not likely to come again." Saltonstall replied to Benton in the negative, discounting the possibility of "regulation by statute."[179]

Undeterred, Benton's enthusiasm shone through a memo to his aide John Howe, in which he shared that he had been rereading NARTB bulletins from 1951.[180] He wrote:

These bulletins, re-read today, sound hysterical—but that's the way the industry has been throughout. And the industry has won throughout. It has terrified the FCC with material of this kind, literally scared them to death, and of course this helps account for the fact that the FCC is the weakest and most venal and terrified of all our Washington agencies—and why I call it a "cartel court."

Outside of the official setting of a committee hearing, and far removed from his political career, Benton was free to air his grievances about the regulatory agency he once treated with respect. He was also increasingly motivated to reach out to as many influential contacts as possible. He sent Senator Warren G. Magnuson (D-WA) a rather long letter asking for help from his committee.[181] Requesting an investigation into the marginalization of educational television stations by the FCC, Benton asked for Magnuson's "leadership," which Benton hoped would be even more impactful given Magnuson's financial ties to the industry (which Benton plainly noted in his letter). The flurry of letters abruptly halted when Benton wrote to Howe revealing that Senator Almer Monroney (D-OK), who "laugh[ed] at the letter to Eisenhower and [wouldn't] even bother to look at it," said Eisenhower's reply was "staff written"—a polite form letter signaling that Benton never had the president's attention.

Conclusion

Because no industry reform resulted from the quiz show scandals, the decade closed with the remains of hopeful reforms strewn about in tatters. Noncommercial and educational television had emerged from the license freeze, however, with a victory. As Perlman points out, educational TV won in the 1950s in a way educational radio did not in the 1930s.[182] With 242 channel reservations, educational TV advocates succeeded where the lone Senator Benton, inexpertly navigating the industry-FCC relationship, had failed. Crusading congressmen may have dragged out the sins of television in its infancy to get some publicity—Representative Lane's abbreviated campaign against TV feels like just such a ploy—but Benton's proposals did not hinge on the "hip swinging torch singers" that Lane railed against. He rooted his reforms in the disruption of a system of broadcasting stagnating under the charge of the same men he worked with as an advertising executive.

Disruption may be a strong word, however, since subscription TV, as Benton stated repeatedly, would be a profitable enterprise. However, the mere mention of a competitor to the commercial system, be it in terms of distribution or content, rattled the industry. Allocating channels to educators was, for commercial broadcasters, a waste of the spectrum; only the entrenched players had the experience to make a go of it. The NARTB's argument did not prevent the JCET from achieving its victory, but by taking a different strategy with Benton the NARTB found its own success. Benton had woven his educational TV advocacy into his plan for the CAB, so the NARTB was able to extinguish the bigger goal by discrediting the smaller one. By returning to their First Amendment toolkit, the association could conflate the commercial system with democratic ideals, casting suspicion on an advisory board that had no discernible checks on its power. As a bonus for the association, advocacy for subscription TV could be swept aside too.

The FCC could not stay out of this fight. Dragged into a controversy of someone else's making, the commission was forced to weigh its commitment to the public interest and programming standards against the zeal of a senator whose intentions were good but whose plans compromised both the industrial and regulatory status quo. Benton's political clumsiness implicated the FCC in what the NARTB perceived to be a regulatory coup of sorts, so the association mobilized and attacked. Either through inertia or capture, the FCC found itself doing its best *not* to endanger the industry's ability to look after itself. For the NARTB, the easiest way to quash the real reform was to bandage the superficial concerns: necklines and crime. In the end, those bandages strangled policy recommendations that, in Benton's mind, could have dislodged television from the commercial radio model.

CONCLUSION

After the Code

The constantly developing processes in and around television threatened to render the Code irrelevant from one year to the next. Aware of the speed of change, the Television Code Review Board (TCRB) was ready to revise the document regularly to keep it alive and responsive to developments, and revise it they did. The Television Code adapted to changes twenty-one different times. The Code was not undone in the early 1980s because it was "outmoded," as John Semonche claims.[1] The NAB scrapped it because its standards had finally crossed a line. The NAB had placed tighter restrictions on ad time during children's programming in the 1970s, a regulatory move completely aligned with the FCC's strong stance against the overcommercialization of programs targeting youth audiences.[2] In 1979, the Department of Justice assessed that the association's limitations on all advertising time—not just those ads during children's programming—artificially inflated the cost of those ads. Not willing to support its own standards in a trial, the NAB agreed to Federal District Court judge Harold Greene's consent decree and killed the Code. The end of the Code proved that even though it was forged in the image and interest of the established commercial model of broadcasting, it could be destroyed by that very same model.

Brooks Hull argues that if the case had gone to trial, Judge Greene would have found in favor of the NAB.[3] His assertion stems from one possible reason for the creation of the Code not addressed thus far. I have argued that ensuring the survival and absolute dominance of the commercial broadcasting industry was the primary catalyst for the Code, and the systematic exclusion of alternatives to that industry was essential to that effort. The Code was one way to retain the status quo. The prevailing understanding of the Code's origins places the NARTB in the position of defending the industry from government intervention. The NARTB publicized the Code as an ethical decision that affirmed broadcasting's role in a thriving democracy.

Although Hull acknowledges the NARTB's defensive role and stated ethical obligations, he posits that the Code might have been an illegal maneuver undertaken to "increase station profit directly."[4] He explains that by limiting commercial time, thus increasing its value to stations, the Code mimicked a type of collusive behavior in which "firms restrict output in an effort to increase industry profit."[5] But because the NAB eliminated the Code to avoid going to trial, no evidence about the Code's financial impact emerged. Hull's study of the Code's financial impact found no evidence that stations profited directly from the Code's strictures. What he did find was that the advertising restrictions, if adhered to, would have improved the viewer experience and benefited advertisers by reducing the clutter that results from "overcommercialization."[6] Because the Code's enforcement mechanism was weak, these ad restrictions mostly went unheeded.[7] So the advertising section of the Code had the potential to please viewers, parents of small children, and advertisers, but as the regulatory climate shifted, that section was the Code's undoing.

The Code's death confirmed its configuration as a living system, a design that presumably safeguarded it at any given time from durable criticisms. It would change with the times, deleting, reforming, and introducing standards for both programs and advertising (as though the two can really be kept separate). The legacy of the Code is tied to its standards for programs; less explored are its standards for ads—an oversight I hope to see rectified. To conclude this study of the run-up to the Code and gauge its reception upon implementation, this final chapter will consider the Code from the perspectives of the people who created it, the people tasked with enforcing it, and, briefly, the people who had no use for it.

Reactions to the Code: Positive, Negative, and Confused

Once the Code went into effect in March 1952, *Television Digest* read the immediate response positively, predicting "unanimous" industry support in short order.[8] Supporting its upbeat tone was data from the NARTB showing a jump in its membership with all four networks back on the roster. *Television Digest* subsequently reported that some advertisers and ad agencies were troubled by the Code's limitations on commercial time, which to them were "intolerable and discriminatory."[9] These were serious accusations that Justin Miller was careful to sidestep.

The ACLU called the document "stultifying and illegal censorship" and believed it "must be abrogated in its entirety."[10] In a letter to FCC chairman

Paul Walker, the chairman of the ACLU's Radio Committee and the executive director of the ACLU wrote that the Code's treatment of "issues of a serious nature" would reduce television to "a diluted force in our national life."[11] For the ACLU, the Code was unconstitutional; only the FCC had the authority to touch programming.[12] In response to this criticism, Justin Miller, at this time chairman of the board of the NARTB, called the ACLU "a bunch of sneaking hypocrites" who were flirting with "communism, fascism, nazi-ism and all other forms of absolute totalitarian government."[13] Sitting down with the ACLU to talk through their issues was therefore an unpalatable option; Miller characterized any such meeting as "a frustrating and unprofitable journey."

Some advertisers seeing the Code for the first time reacted positively, and their reactions were recorded in *Sponsor*, a trade magazine serving the advertising community. A commentary on the Code written by Norman R. Glenn, editor and president of the magazine, included praise from the president of an unnamed agency, as well as from executives at J. Walter Thompson and Biow, but one response from a sales executive is a telling example of general befuddlement: "Don't ask us. We don't know. Tell us what to think."[14] Glenn pointed to this reaction as emblematic of the surprise that the Code might cause. He believed that the Code had teeth, calling it a "robust lion's roar compared to the earlier mouse-like squeaks" of the radio codes.[15] What stood out about the Code to Glenn were the complaint and punishment mechanisms, as well as the presence of the Seal of Good Practice, which signified "respectability."[16] Not usable on radio, the Seal functioned as a visual representation of compliance and, above all, accountability to the public.

Glenn's commentary also cited "skeptics" who were doubtful that local broadcasters would be impressed by the symbolism and opt for ethics over ratings.[17] Miller referenced these skeptics in a letter to Glenn following the publication of his piece. Miller praised Glenn's analysis of the Code and thanked him for elevating the discourse about television in general. More importantly, Miller pointed to the criticism heaped on the Code by others in the popular and trade press, referring to their skepticism as "catastrophic."[18] Not surprisingly, Miller carried on a cordial relationship with Glenn, as any head of the NAB would need to. Miller wrote to him the following month to thank him for sending along a photograph of the Television Code Committee. He wrote, "It goes into my collection [. . .] to commemorate another milestone in television history."[19] Miller had presided over television's inclusion in the NAB, and he was now seeing television transition into a new, presumably more orderly phase. Peace did not come, however, with the drafting and implementation of the Code.

Testing the Code: Congress Investigates TV

Just over two months after the implementation of the Television Code, the Federal Communications Commission Subcommittee of the Committee on Interstate and Foreign Commerce in the House of Representatives, under the direction of H. Res. 278, opened hearings into radio and television programming. In the best cases, congressional hearings create meaningful spaces in which industry and non-industry groups alike can speak to circumstances triggering or already addressed by policy.[20] This particular subcommittee had within its purview the revision of the Communications Act of 1934, so these hearings were not without potential consequence.[21] Congressman E. C. Gathings's (D-AK) inciting resolution set its sights on the presence of "immoral or otherwise offensive matter" on radio and television, with a particular emphasis on indecent and violent programs.[22] Upon conclusion of its investigation, the subcommittee would have the discretion to recommend legislation to clean up TV.[23] Chaired by Representative Oren Harris (D-AK), the subcommittee questioned witnesses from the broadcasting industry, the NARTB, the advertising industry, and various citizens' groups. The networks did not want to take part, preferring to allow the NARTB and its Code leadership to speak for the industry.[24] CBS executives appealed to Representative Harris to omit the networks from the roster of witnesses, but the chairman insisted that they be present.

The hearings lasted from June through December 1952, with testimony broken up into three categories: the presence of "immoral or otherwise offensive matter" on radio and television; the "emphasis [. . .] upon crime, violence, and corruption"; and the industry's self-regulating actions.[25] Harris clarified that the hearings would not seek to resolve the debate over what constituted the "most desirable types of programs."[26] Additionally, it did not appear that the hearings sought viable alternatives to the industry's existing plan of attack. In fact, the composition of the witness list privileged commercial stakeholders, resulting in an unsurprising affirmation of the industry's behavior. Of the thirty-two witnesses appearing before the subcommittee, fourteen represented commercial broadcasters and advertisers, eight represented religious or temperance groups, and three were congressmen. According to its report, the subcommittee knew that other witnesses would have broadened the conversation, but time was limited. So while multiple religious organizations testified, noncommercial educational broadcasting advocates were scarce. Speaking for educational television was Alice Keith, chairman of the National Academy of Broadcasting Foundation Board, a nonprofit organization dedicated to bringing together commercial television and educational programming.[27]

Not surprisingly, witnesses spun educational television into a commercial success story, noting that such programming was having an easier time attracting viewers and sponsors.[28]

Prior to the hearings, several members of the NARTB leadership met with the subcommittee to form a game plan.[29] The networks and the association pled similar cases, both asking for leniency in consideration of the experimental phase of television's development and the medium's "voracious" appetite for content.[30] The networks craftily described their own helplessness within the situation they fought to create and maintain. "The television networks carry the brunt of the program burden," the report stated, emphasizing stations' dependence on national programming but not, of course, how the networks had used live programming to manufacture that dependence.[31]

The industry defended itself capably and found fault with others. One particular witness placed blame on the shoulders of writers and performers who "were recruited from the cabarets and nightclubs of the large metropolitan areas" and had not yet learned how to tame their skills for domestic consumption.[32] Industry witnesses also blamed viewers for the state of things, insisting that the industry itself had actually tempered audiences' lust for crime programming. According to the subcommittee's report, "It was alleged that were the stations to give in to the public's demand for [crime programs], there would be many more than there are now."[33] As it stood, crime programs made up only 2.7 percent of NBC's schedule, according to Charles Denny's testimony.[34] Recall that one of the viewers' main complaints about crime programming was scheduling: too often these programs aired when children were awake and watching television. The networks contradicted these complaints, testifying that crime shows did not appear before 9:00 PM. Pro-industry witnesses also countered the argument that westerns negatively affected children by offering expert opinions that such programs "satisfied a real need on the part of juveniles in providing a harmless outlet for their aggressive tendencies."[35] They also testified to the Code's success by mentioning the decline in the number of complaint letters focusing on indecent costuming.[36]

Religious and temperance organizations ostensibly spoke on behalf of viewers, but the public was not as well represented as the industry was. The subcommittee had access to 1,200 "letters, petitions, and other communications," which we know from chapter 4 were mostly supplied by the FCC. Oddly, the subcommittee's report did not refer to that feedback, preferring instead to highlight the industry's testimony dismissing national "write-in campaigns" and asserting that the networks, the stations, and the NARTB "had received only few communications from individuals pointing out specific instances of alleged immorality."[37] The industry predictably minimized public criticism, but the subcommittee's report concluded that an "alert, articulate public" needed to be in constant conversation with "an alert and conscientious broadcasting industry."[38] The industry needed a watchdog on the ground, not just in Washington, and the subcommittee believed that viewers needed to have a greater presence in this regard.

The NARTB took the subcommittee's final report very well, believing that it was "favorable to the Code" since the document's "spirit, purpose and intent pointed to improvement in television programming in the brief time the Code had been in effect."[39] Although the subcommittee concluded that broadcasters, in fact, aired offensive programs and an excess of crime programs, the report regarded the Code as a positive self-regulatory effort, evidence of "the sincerity of the industry."[40] Truly, the subcommittee felt that government intervention in the area of content, while possibly justifiable, had "potential evils [that] might be even greater than the evils that such controls might be designed to remedy."[41] However, the report was careful to note that the Code should not interfere with the licensor-licensee relationship; licensees could not consider the Code to be the ultimate arbiter of their commitment to the public interest.[42] This reaffirmation of the authority of the FCC, while undercutting the usefulness of the Code somewhat, also appeared to erect a wall between state and private action.

At the end of the report, the subcommittee recommended that its investigation be continued into the next Congress in order to maintain a steady surveillance of the industry's self-regulatory behavior. This did not come to pass, but the threat of a watchful congressional presence lingered as the Television Code Review Board standardized its own practices and clamped down on Code violations. The industry had achieved its goal, however. The substantial industry presence in the hearings was able, in the opinion of the chairman of the TCRB, to steer the subcommittee "to assume a most reasonable and welcome attitude toward the industry."[43] The NARTB was free to go about its business of persuading the nation that TV was safe for consumption.

Enforcing the Code: The TCRB and Its *First Report to the People of the United States*

Committed to the idea that it was answerable to the people, as though it were a body of elected officials, the TCRB issued its first-year review in November 1953. Harold Fellows characterized the TCRB as a trailblazing self-regulatory body, and certainly it was the most organized enforcement body the association had ever seen. The TCRB was composed of five local television executives.

John E. Fetzer of WKZO-TV in Kalamazoo, Michigan, was the chairman; J. Leonard Reinsch, managing director of WSB-TV in Atlanta, was the vice chairman; and E. K. Jett, vice president and director of television for WMAR-TV in Baltimore, was a member. Notably, the TCRB included one woman. Dorothy Bullitt née Stimson, owner of KING-TV in Seattle, hailed from a prominent Seattle family and had been educated in New York.[44] Her husband worked with her family in their real estate business, and tragically, in the space of three years, she lost her father, brother, and husband. In their absence, she assumed control of the real estate business and flourished in other ways as well. A delegate to the Democratic National Convention and a member of the Washington Emergency Relief Commission as well as the National Women's Committee for the Mobilization for Human Needs, Bullitt established herself in the 1930s as a politically active humanitarian. In the 1940s, she entered the radio business, purchasing a radio station and starting her own broadcasting company. And in 1949, her company bought a television station. Despite this hefty résumé, *Broadcasting* succinctly described Mrs. A. Scott Bullitt as "the mother of three children."[45] Her appointment to the TCRB was, according to the trade magazine, "an obvious gesture to bring into the code's operation the family and home viewpoint, giving balance to the structure." The final member of the TCRB was Walter J. Damm of WTMJ, a station executive (and president of the NAB from 1930 to 1931) whose influence in broadcasting, woven throughout this book, helped to nudge the NAB in television's direction.

Complete with procedures for Seal revocation and appeals, the TCRB, under Fetzer's leadership, emerged from year one with a number of successes to its credit. One sign of success, by the NARTB's metrics, was the presence of the four television networks on the roster of Code subscribers. The association's tumultuous relationship with ABC and CBS, as discussed in chapter 2, had threatened to derail the widespread adoption of the Code. However, by the time of the *First Report*'s publication, 188 stations (out of 299 total operating stations) and all four networks had subscribed.[46]

Another sign of success was the TCRB's record of adjudicating complaints. A description of the schedule of the director of Television Code affairs reveals the amount of time devoted to minding the store: he "spends approximately one third of his time in the field visiting subscriber, and non-subscriber, stations and regularly checking in at the different network headquarters where much time is spent with continuity acceptance editors and occasionally show producers or directors."[47] The *First Report* offered examples of situations in which viewer complaints were resolved through the cooperative efforts of networks, the TCRB, and stations. From a complaint about costuming to another about a program's representation of a suicide, to a letter about an off-color

FIRST REPORT

To The People of The United States

Television Code Review Board

National Association of Radio and Television Broadcasters

1771 N Street, N. W.

November, 1953   Washington 6, D. C.

FIGURE 8.1. NARTB Television Code Review Board's *First Report to the People of the United States* (1953)

The *First Report to the People of the United States* described the successes of the Television Code Review Board and the value of viewer complaints. Courtesy of Wisconsin Historical Society, WHS-133413.

joke—all resolved—the TCRB assured readers that the mechanisms put into place by the Code were efficiently and satisfactorily stemming the tide of offensive programming.

The *First Report* included letters from optimistic viewers, as well as comments from members of the 83rd Congress, solicited in early 1953 by Chairman Fetzer. The unnamed senators provided mainly positive comments, expressing similar sentiments regarding the intentions of the Code. Some senators who had not yet read the Code were neutral, and one eager senator fancied himself an industry professional, remarking, "I will be happy to send to your Washington office any television programming suggestions which I consider to be constructive."[48] Comments from the House tended to be more critical, with one warning the TCRB outright that failure to "control those who would exploit" viewers would result in legislation "not [. . .] to your liking."[49]

The *First Report* promoted TCRB successes in terms of Code subscriber numbers, complaint resolutions, and Code adherence. A survey conducted by the TCRB in the spring of 1953 determined that, "while some defections" occurred, "subscribers were following the Code successfully."[50] Whether the first year's results were "the culmination of sincere thinking and action by an industry composed of men and women of high moral integrity who recognize that service to the people of this country is the industry's greatest responsibility," as the *First Report* stated, is debatable, but we know for certain that the results represented a concerted effort to prove that the commercial system was the one best equipped to service the public.[51] However, the report was the only public relations effort of its kind. According to Morgan, the *First Report* was "favorably received" and successful in its attempt to raise awareness of the Code.[52] The NARTB planned to distribute part 2 of the report early in 1954, but for reasons difficult to determine, the TCRB declined to follow up with subsequent reports.[53]

Chairman Fetzer spoke to attendees of the NARTB's annual convention in 1953 and described the tightrope that the Code had walked in its first year. The TCRB had encountered the same problem the NAB had wrestled with in each of its radio codes. Stations subscribed to the Code *voluntarily*, so to enforce it with no significant aftershocks was difficult. The TCRB could, in Fetzer's words "wield the big stick [. . .] but if the net result should bring about mass resignations from the Code, or bring litigation on the basis of restraint of trade, or any one of a number of reversals, the whole Code could collapse before we have it off the ground."[54] The TCRB had to steer itself through complicated legal and industrial surroundings, but in his speech Fetzer maintained that the board had plenty of copilots. One network employed "special monitors" to "spot-check" programs' compliance with the Code. Another network enlisted its Continuity Acceptance Department to align its practices with the "spirit of the Code." At a third network, the Continuity Acceptance Department reviewed all products, scripts, motion pictures, and advertisements before and after airing, and at the fourth network regular bulletins notified employees of changes made to programs in accordance with Code standards. This crackdown on content, whether exaggerated by the networks for the benefit of the TCRB or an earnest attempt to stave off external criticism, was made visible through the Code's Seal of Good Practice.

Promoting the Code: The Seal of Good Practice

Baldwin, Cave, and Lodge write that "qualifications, certificates, or marks of quality" all function as "incentives" for companies to take part in voluntary self-regulation.[55] The purpose of such a certificate or mark seems obvious, but in the case of the NARTB's Seal of Good Practice, the amount of energy funneled into the creation and maintenance of that sort of signifier speaks to a disconnect between the association's desire for credibility and its member stations' realities. The Seal of Good Practice was ostensibly a symbol of high standards, to be displayed by Code subscribers in good standing with the Television Code Review Board. The TCRB could revoke the Seal if a station violated the provisions of the Code, so its appearance on the screen announced the station's continued adherence to the NARTB's standards. The Seal was, in theory, a vital public relations tool, an ambassador of sorts representing the association to the public through the local station. In practice, the Seal might have fallen flat. Morgan documents that stations gave the Seal little play, ignoring it altogether or throwing it on the screen "during [. . .] low audience periods."[56]

Chairman Fetzer worried that the Seal in its original form was ineffective (see figure 4.2). In a letter to Justin Miller, Fetzer explained that the job of the TCRB was to "sell the seal" to both the industry and the public.[57] Selling it to the industry was not difficult, but from a design perspective, and with the public in mind, the Seal "misse[d] the point." Fetzer maintained that the placement of "NARTB" so prominently was "just as confusing to the public as it originally was to members of the industry" when the NAB changed its name. The point was not to publicize the name of the association; rather, the aim should have been to inform the public (or "sell them on the fact") that the TCRB existed at all. Emblazoning the name of the association on TV screens was counterproductive because few people knew what trade associations did. Trade associations had a reputation for being "anti-something-or-other," so publicizing the association over the Television Code Review Board made no sense to Fetzer. The revised design proposed for the Seal in June 1952 replaced "NARTB" with "Television Code Board Washington, DC."

Fetzer also wanted the Seal to direct viewers to send letters to the TCRB rather than to the government. Calling such a solicitation a "godsend" for Congress, the FCC, stations, and networks, Fetzer wrote, "Any device which we can erect to take the heat off all these instrumentalities and place the burden on the Television Code Board should be welcome." Fetzer assumed that stations and networks, especially, would happily direct complaints to a centralized, self-regulatory repository. The opposite was the case, however. Morgan writes that the stations downplayed the Seal because they wanted viewers to write directly to stations, not to the TCRB.[58] Their reaction was perfectly rational. While the TCRB saw itself as helping stations and networks by channeling viewer complaints through the trade association, thus bypassing the government and reducing regulators' and lawmakers' exposure to complaints, stations must have seen the TCRB as yet another authority that could come between their business and their license to operate that business. Morgan notes that when a revised Seal debuted in 1955, the Washington address remained in place.[59]

Justin Miller responded positively to Fetzer's suggestions to boost the strength of the Seal, and he further recommended that the revised Seal move the word "Information" closer to the word "Entertainment."[60] Doing so would align the Seal with the association's progress in persuading critics that broadcasting was worthy of First Amendment protection. But "NARTB" needed to stay in the Seal, even if it was moved to the bottom, so that the emblem relayed the "full context" of the Code. Miller hinted that the association's name did not need to remain visible forever. The strong possibility of antitrust violations resulting from the Code's rules and enforcement might even necessitate the "full separation of the Code Board and NARTB" at a later date.

The public's ignorance of the value of the Seal and the Code persisted regardless of where or if "NARTB" appeared on the screen. According to Morgan, stations were reluctant to waste screen time on the Seal if it had no effect on the viewer at all.[61] Many felt the solution was to overhaul the TCRB's promotion. The need to revamp the Code's image was apparent to organizations other than the stations. A representative from the Dancer-Fitzgerald-Sample advertising agency described the Seal display as a weak gesture: "The only thing the Association and the stations do now is put a shield a couple of times a day in bad time periods and say 'we're a member of the National Association of Radio and Television Broadcasters and we're good boys—we have a code, etc.'"[62] The advertiser suggested displaying the Seal "for a second" but removing it "in a hurry" and then cutting to a spot with young people professing their excitement for an upcoming television program. The advertiser's advice raised a question neglected by the entire Seal discussion: how did the symbol of quality relate, if at all, to the actual quality of the upcoming television programs?

Missing from the Seal—Good TV: TV Writers on the Code

The report issued at the conclusion of the 1952 Gathings hearings noted that one witness faulted writers for objectionable material since a sizable proportion of the writers hailed from less than reputable entertainment venues in big cities.[63] No writers appeared on the witness list at the hearings, so no one spoke in their defense. But in 1953, two screenwriters seized the opportunity to express their displeasure with the Code's strictures via the Radio Committee of the ACLU. For Mort Lewis (*Suspense*) and Morton Wishengrad (*With These Hands*), the Code fostered programs that would "present life unrealistically," an indicator that the document thought little of television viewers.[64] By contrast, the writers had "confidence in the intelligence of the American people" and wished to see certain Code restrictions lifted in order "to raise the standards of this great medium to a higher level."

Lewis and Wishengrad first tackled the provision prohibiting the representation of divorce "as a solution for marital problems" (see appendix A). The writers took issue with the "immorality" of forcing television wives to stay married to abusive husbands or television husbands to stay with wives who abandon them. To assert otherwise was to concoct a "dream world" for the viewer. Equally unrealistic was the Code's desire to portray all law enforcement officials as upstanding. Lewis and Wishengrad asserted that the police were "no better or worse than we" and so should have a presence on the screen that allows for "the crooked sheriff" or any other morally problematic officer. Furthermore, what would writers be forced to change to adhere to the rule forbidding suicide? "Instead of [Judas Iscariot] hanging himself, is he to be impaled by a Roman soldier, or what?" the writers asked.

Lewis and Wishengrad also argued against Code provisions restricting the manner of presentation, particularly in terms of tone. The prohibition on stylistic effects that "would shock or alarm the viewer" would mean that footage of the atomic bomb exploding or films such as Roberto Rossellini's *Open City* would have no place on television. For the writers, this ban was "meaningless" and would surely "bring the entire code into disrepute." Additionally, the unacceptability of "frivolous, cynical, or callous" treatments of crime struck the writers as "impossible" to implement. "Very few hold-up men are gentlemanly about it," they retorted.

A Few More Lapses, Then—Censorship

By FAYE EMERSON.

Censorship has always been an ugly word in America. It implies a lack of freedom, a dictation by superior beings over what we shall see, hear, or read. And being Americans and independent, we have kept censorship to the minimum required for decency and to protect the morals of our young.

The great television networks are very sensitive to the problems created by their enormous coverage, and by the fact that they go directly into the homes where the audience may consist of anybody and everybody from Grandma down to Little Suzy, aged 2.

Faye Emerson

Long ago the television bigwigs got together and drew up their own code. And I must say I think they've done a pretty good job of sticking to it. There is a lot of blue material that gets past, but it's usually double entendre or ad lib stuff on panel shows, which is pretty hard to regulate.

* * *

All this brings me to the point of discussing Betty Hutton and NBC-TV's first spectacular "Satins and Spurs." Everyone in this business knows that Max Liebman, the producer, is a man of taste and ability, who consistently turned out great entertainment over the years with Sid Caesar and Imogene Coca. So it's hard to imagine how he could allow two such regrettable lapses of taste as those which occurred in "Satins and Spurs."

One was when she spat a mouthful of food and milk right at the camera—and in color, too. And the second was a little production number entitled "Sexy Sadie," which would have made even Howard Hughes think twice. Mind you, this all occurred early on a Sunday evening when all the family, including the small fry, were gathered around to watch that great event.

I can well imagine that there was fairly heavy mail from parents who have been trying to convince their offspring that ladies and gentlemen definitely do not spit food on the floor. Perhaps the "Sexy Sadie" number went over the heads of small children, but it wasn't exactly subtle. Also it was completely extraneous to the plot, so it seemed to be sex for sweet sex alone.

Enough lapses of this kind will inevitably bring on real censorship from outside the industry, with all the unpleasantness that the word implies.

I don't mean to suggest that television must be sterile and immature in its approach to adult problems or adult humor, but it doesn't have to be burlesque either.

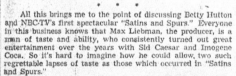

RECEIVED

SEP 21 1954

SYLVESTER L. WEAVER, JR.

FIGURE 8.2. David Sarnoff note to Pat Weaver (1954)
Responding to more criticism of television programming well after the implementation of the Television Code, RCA chairman David Sarnoff asked NBC executive Pat Weaver if management did not share some of the blame for problematic content. Courtesy of Wisconsin Historical Society, WHS-133408. Image courtesy of NBCUniversal Media, LLC.

In their 1951 book entitled *The Television Progam: Its Writing, Direction, and Production*, Edward Stasheff and Rudy Bretz devote a few pages to the task of working within the constraints of industry codes. They write, "The novice may ask how he is to write meaningful drama within these rigid limitations; the veteran, while regretting the necessity, realizes that he can still tell a few thousand worth-while stories without overstepping the bounds self-imposed by the industry."[65] Lewis and Wishengrad, veteran writers, could not temper or suppress their objections so easily. They conceived of good television as programs and films that did not shy away from the darker episodes of life, and their brief thoughts on the subject, as relayed to the NARTB, clashed with the association's view, mainly in that the association had formulated *no* views as to what made for good TV. The work of the trade association had little connection to the daily creation of dramas, comedies, soap operas, or variety shows. The words of Lewis and Wishengrad highlighted that disconnect and gave voice to the creative workers excluded from Code deliberations and committee hearings.

A Welcome Guest in the House

The four words on the original Seal of Good Practice—entertainment, education, culture, and information—indicated intent but not execution. These were buzzwords assigned to bring television in line with protected media. The NARTB's challenge was to translate its pithy branding into a meaningful message for its stations. To that end, the NARTB pulled more marketing tools out of its toolbox. Well after the Television Code was operational, the TCRB eventually assembled a promotional package that "included a library of slide and sound film spot announcements ranging from ten seconds to one minute in length, as well as reproduction of the seal for use in ads or stationery and brochures for distribution to the public."[66] That effort, completed in 1955, had little positive effect on the distance between the public and the Code, so a second promotional push launched in 1957 under the theme "You see better TV on a Television Code station."[67] Included in this campaign was a short propaganda film entitled *A Welcome Guest in the House*.[68] Produced in 1957 by Westinghouse and the NARTB, *A Welcome Guest in the House* drove home the association's feelings about government by focusing on a young, blond-haired, middle-class boy from "Your Street, U.S.A." The film delivered the meaningful, ideological message the NARTB needed to clarify its role, duty, and power.

In the first shot of the film, the boy wanders through a field of tall grass and weeds, and an abrupt cut reveals a low-angle shot of a television transmission tower, the technology that facilitates the young boy's relationship with his television receiver. The film's narrator, Don Ameche, explains that the boy, who represents "the future of our country," "spends a lot of his time in the company of a steel tower." The boy also represents "a challenge to men everywhere who believe in freedom and truth." These men are broadcasters "who distribute ideas and strengthen ideals." More shots of the boy playing outside are intercut with images of various television antennas. The final image of an antenna dissolves into a shot of railroad tracks, along which our boy ambles. These exterior shots establish a strict dichotomy between an old means of transport (the railroad) and a new, invasive one (the television).

In the next segment of the film, before which the narrator explains that the "welcome guest" in our homes is the NAB, the boy enters his house and basks in the "beam" of a television screen.[69] This tight shot of the boy, lit only by the television set, is the single discernible image on the screen; the rest of the frame is bathed in complete darkness. This "beam," as the narrator clarifies, allows the boy to "learn of forces at large alien to his way of life." At this point, the film cuts between the boy's glowing face and stock footage of marching Soviet troops. Because "his horizons are no longer confined to the front yard," the modern information age enabled by television requires a watchful eye, an official minder. The narrator details how the Television Code ensures the responsible "treatment of news and public events." From the global—a revolution in Hungary, the sinking of the *Andrea Doria*—to the local—a tornado, a jury trial in Texas, a prison riot in Utah, trapped miners in Pennsylvania— responsible broadcasting is illustrated through the use of news footage. The narrator also connects three broadcasts—the tornado coverage, as well as a documentary on urban squalor and renewal and another on traffic congestion—to tangible change: the ensuing relief for tornado victims and legislation regarding urban spaces and traffic.

The mood of the film lightens once the narrator discusses television's value as "the best seat in the house" for sporting events, royal coronations, and political conventions. The narration makes clear that the medium that delivers these events is "free commercial television," not pay television or noncommercial educational television. A quick detour to Hollywood and the Emmy Awards brings the film back to what television is known (and criticized) for: its entertainment values. As the narrator explains, "Yes, our medium is an entertainment medium, a commercial advertising and free entertainment medium. And it's all the stars and all of the shows of the networks, affiliates, and independent stations that make possible this exciting and wonderful gift called television."

A serious tone returns to the film as somber music and another shot of a transmission tower displace the upbeat music and shots of smiling celebrities.

This brief segment of the film, which lasts no longer than one minute, addresses religion and spirituality, two responsibilities of the broadcaster under the Television Code. The conclusion of the segment positions the NAB's Seal of Good Practice as a tool of religiosity. Light shines through a stained-glass window and passes through the lower left-hand corner of the Seal.

The valorization of church transitions to a rejection of the state as the film cuts to its final major segment situated around NAB headquarters in Washington, DC,

> the home office of your broadcasters' association governing body of broadcasting stations across the land—stations who have voluntarily banded together to assure the American television public that your television set will always be a welcome guest in the house. It is here that the Code Board of Review meets to determine policy and enforcement of high broadcasting standards. Monitor reports are scrutinized and appropriate action is taken. A broadcaster might win an accolade for his efforts. Or for his negligence it could be as severe as expulsion. In this regard, the governing body is similar to that which we have had in other mediums. Always a self-regulatory system which functions without government censorship—a system which, in this country, we have found works very well.

As discussed throughout this book, NAB officials openly and consistently frowned upon government interference in programming. The fact that the film never mentions the FCC or even Congress demonstrates a willingness to erase the federal government from the narrative of television oversight. The TBA did something similar in its heyday. As the "governing body" in Washington, DC, the NAB simply replaces the FCC in *A Welcome Guest in the House*.

The film eventually takes us inside NAB headquarters, where the five members of the TCRB are assembled around a conference table. In a subsequent montage, TCRB members point at graphs and charts and reassemble to confer with each other. Next, a shot of a professional man and woman, once again lit by a television screen, stare, point, and take notes. Another man is stationed in front of not one but three monitors, viewing three different programs. The narrator explains, "Implementing [the TCRB's] decisions is a permanent staff which keeps abreast of broadcasting practices. Perhaps it sounds like a lot of administrative routine behind all the glitter. But the truth is, these people are not any different than you and I." The TCRB segment of *A Welcome Guest in the House* attempted to ease any fears of authoritarian surveillance that broadcasters might have harbored. At the same time, shots of five TCRB members sitting around a conference table in Washington,

DC, did little to dislodge the possible fear of being judged by a centralized nongovernmental body.

A brief mention of television in the classroom cuts quickly to our boy leaning against a car in a shopping center parking lot. Over a tracking shot of all of the parked cars, the narrator locates television within the ideology of the free market and the ideals of democracy:

> The free television system is based on the competitive flow of ideas, which, in turn, contributes to a flow of goods. The economists tell us this is the basis of free enterprise and in turn is fundamental to a democratic system. Our way of life, then, is based on standards of quality and taste even though appetites change. Our appetites for television change, too. And the Code Board of Review depends upon listener reaction for criteria in standards. Your letters are a great help in this direction. Letters of comment on advertising practices, programming material, and suitability of performers, are all a part of Code Board of Review decisions. Your letters contribute to the American system of self-regulation, reflected by newspapers, periodicals, motion pictures, and the theatrical arts.

The nods to capitalism, democracy, viewer feedback, and self-regulation, all neatly compressed into one monologue, represent the exceedingly rigid narrative the NARTB created for itself. The priority was to sustain the commercial system, which, the film tells us, was necessary for a functioning democracy. Democracy revealed itself through the input of viewers, who helped to tighten controls on television. Viewer control and NARTB control were preferable to government control.

Finally, the monologue attempts to place television within the protected space occupied by newspapers, magazines, movies, and the theater, once again erasing institutional history. Before the Radio Act of 1927, the Hoover conferences affirmed that radio would be a medium regulated by the government. As an over-the-air medium, television joined radio within the prevailing regulatory framework. In the universe of the film, however, the only legitimate governing body was the trade association.

Morgan explains that *A Welcome Guest in the House* "was intended to be supplemented with a five-minute cut-in, in which the local station would explain what it was doing to implement the Code."[70] The choppiness of the film appears to welcome that sort of locally produced supplementary material. Like the earlier TCRB campaign, this one suffered from a poor communication strategy. Morgan quotes one viewer who recognized that most of the film focused on the idea of television rather than the enforcement of standards.[71]

In the end, only twenty-three stations featured the film in their schedules.[72] The national narrative got lost in the local translations.

Defending the Code

By 1954, two years after the implementation of the Code, Fetzer offered his assessment of the document in an address before the NARTB in Chicago. At that time, the Code counted among its subscribers 225 member stations (out of 380 total operating stations) and all four networks, although 30 NARTB members had refused to subscribe.[73] During the Code's brief existence, the TCRB had "received some 600 comments upon television programming and advertising practices," with most being complaints, "as was to be expected." One marker of success for Fetzer, given the urgency to avoid provoking the FCC, was the decreased number of complaint letters sent to the government regulator. According to Fetzer, the FCC received 967 complaint letters in an eleven-week period in 1951. In 1953, "over a similar eight-week period," the FCC received 144 complaint letters. Fetzer lauded the efficiency and coopera-tion of the TCRB and its staff in responding to and acting on the complaints "when necessary." He also pointed out that stations were under no duress at any point. When the TCRB monitored top programs on member stations clandestinely, the monitoring reports would sometimes require action on the part of stations that had strayed from the Code's standards. Fetzer assured the audience, "It is fundamental to the voluntary policy of this self-regulatory procedure, to tell you that on the basis of the monitoring reports alone, the stations and networks took voluntary corrective action." Fetzer also assured advertisers that they were a crucial component of this "enlightened program of self-regulation." Another essential part was the Seal of Good Practice, which Fetzer lamented was only displayed by many stations in the morning and at sign-off. Fetzer expressed frustration that the neglectful stations did not re-gard the Seal with the respect it required. For the TCRB, the Seal was "the best and greatest weapon" they had—a weapon against intervention from both the Senate and the House, as well as from the FCC.

In a 1956 speech, Harold Fellows impressed upon his audience that the Code, already in its third edition, was not a centralized initiative; its mainte-nance depended on the cooperation and participation of local broadcasters.[74] Local compliance was paramount because failure to act on a local level might have led to dangerous action on a national and federal level, according to Fel-lows. He pointed to "government agencies" and "pressure groups [. . .] deter-mined to storm the castles of the wicked and villainous broadcasters." "We may mock them," he said, "but they are not to be under-estimated—none of them." Evidently, the fearmongering present in Miller's speeches years earlier remained within the rhetoric of the NARTB well after the Code was in place. Countering the specter of government involvement was an accentuation of the victories of self-regulation. Fellows congratulated the NARTB for examining its operations with a critical eye and for its continuing willingness to look for its own flaws. He saw this tough self-examination as evidence of the Code's strength and continued growth. "This, basically is your Code," Fellows said to the broadcasters in the audience. "By observing its precepts you are contributing materially to the prosperity of American free enterprise and to the enlightenment and well-being of our citizenry."

Absent from NAB/NARTB speeches is a thoughtful consideration of television's might as an artistic medium or its potential innovations as a storytelling medium. Television's trade association willingly carried broadcasting on its shoulders, but it had little regard for the screen's substance beyond those annoyances that could destabilize its smooth operation. Since the audience is the product in a commercial broadcasting system, it makes sense that the association would sacrifice a concern for innovative or even plain entertaining content in order to pacify vocal, angry consumers. But because the self-sustaining logic of the trade association required the standardization of products and behaviors, content was classified as the product and had to be controlled as such.

Important, too, is the system in which that content was produced and distributed. With the networking of the United States, content crossed state lines from a few networks to hundreds and eventually thousands of stations. So it is imperative that we resist thinking of the association as an anonymous national group with no ties to the basic unit of broadcasting. The association was the local stations, though the larger stations with network affiliations dominated the smaller, independent stations. Through an examination of the trade association, we have been able to tap into the interests of the license holders, since they—not the networks—are answerable to the government. The entire regulatory mechanism depended on the license, an increasingly foreign concept in the face of cable and streaming outlets. In television's infancy, the license was a type of power subject to oversight from the government, beholden to the public interest, vulnerable to exploitation by the networks, and welcomed into the community of the trade association. Under those circumstances television emerged, simultaneously constrained and liberated by a philosophy that insisted on shaping the world around it by excluding all possible alternatives.

The Television Code: Section on "Acceptability of Program Material"[1]

Program materials should enlarge the horizons of the viewer, provide him with wholesome entertainment, afford helpful stimulation, and remind him of the responsibilities which the citizen has towards his society. Furthermore:

a) (i) Profanity, obscenity, smut and vulgarity are forbidden, even when likely to be understood only by part of the audience. From time to time, words which have been acceptable, acquire undesirable meanings, and telecasters should be alert to eliminate such words.

    (ii) The Television Code Review Board . . . shall maintain and issue to subscribers, from time to time, a continuing list of specific words and phrases which should not be used in keeping with this subsection. This list, however, shall not be considered as all-inclusive.

b) (i) Attacks on religion and religious faiths are not allowed.

    (ii) Reverence is to mark any mention of the name of God, His attributes and powers.

    (iii) When religious rites are included in other than religious programs, the rites are accurately presented, and the ministers, priests and rabbis portrayed in their callings are vested with the dignity of their office and under no circumstances are to be held up to ridicule.

c) (i) Contests may not constitute a lottery.

    (ii) Any telecasting designed to "buy" the television audience by requiring it to listen and/or view in hope of reward, rather than for the quality of the program, should be avoided. . . .

d) Respect is maintained for the sanctity of marriage and the value of the home. Divorce is not treated casually nor justified as a solution for marital problems.

e) Illicit sex relations are not treated as commendable.

f) Sex crimes and abnormalities are generally unacceptable as program material.

g) Drunkenness and narcotic addiction are never presented as desirable or prevalent.

h) The administration of illegal drugs will not be displayed.

i) The use of liquor in program content shall be de-emphasized. The consumption of liquor in American life, when not required by the plot or for proper characterization, shall not be shown.

j) The use of gambling devices or scenes necessary to the development of plot or as appropriate background is acceptable only when presented with discretion and in moderation, and in a manner which would not excite interest in, or foster, betting nor be instructional in nature. Telecasts of actual sports programs at which on-the-scene betting is permitted by law should be presented in a manner in keeping with Federal, state and local laws, and should concentrate on the subject as a public sporting event.

k) In reference to physical or mental afflictions and deformities, special precautions must be taken to avoid ridiculing sufferers from similar ailments and offending them or members of their families.

l) Exhibitions of fortune-telling, astrology, phrenology, palm-reading, and numerology are acceptable only when required by a plot or the theme of a program, and then the presentation should be developed in a manner designed not to foster superstition or excite interest or belief in these subjects.

m) Televised drama shall not simulate news or special events in such a way as to mislead or alarm. . . .

n) Legal, medical and other professional advice, diagnosis and treatment will be permitted only in conformity with law and recognized ethical and professional standards.

o) The presentation of cruelty, greed and selfishness as worthy motivations is to be avoided.

p) Unfair exploitation of others for personal gain shall not be presented as praiseworthy.

q) Criminality shall be presented as undesirable and unsympathetic. The condoning of crime and the treatment of the commission of crime in a frivolous, cynical or callous manner is unacceptable.

r) The presentation of techniques of crime in such detail as to invite imitation shall be avoided.

s) The use of horror for its own sake will be eliminated; the use of visual or aural effects which would shock or alarm the viewer, and the detailed presentation of brutality or physical agony by sight or by sound are not permissible.

t) Law enforcement shall be upheld, and the officers of the law are to be portrayed with respect and dignity.

u) The presentation of murder or revenge as a motive for murder shall not be presented as justifiable.

v) Suicide as an acceptable solution for human problems is prohibited.

w) The exposition of sex crimes will be avoided.

x) The appearances or dramatization of persons featured in actual crime news will be permitted only in such light as to aid law enforcement or to report the news event.

The Television Code: Section on "Decency and Decorum in Production"[1]

1. The costuming of all performers shall be within the bounds of propriety, and shall avoid such exposure or such emphasis on anatomical detail as would embarrass or offend home viewers.
2. The movements of dancers, actors, or other performers shall be kept within the bounds of decency, and lewdness and impropriety shall not be suggested in the positions assumed by performers.
3. Camera angles shall avoid such views of performers as to emphasize anatomical details indecently.
4. Racial or nationality types shall not be shown on television in such a manner as to ridicule the race or nationality.
5. The use of locations closely associated with sexual life or with sexual sin must be governed by good taste and delicacy.

Notes

Introduction

1. Semonche, *Censoring Sex*, 194.
2. See Morgan, "The Television Code of the National Association of Broadcasters."
3. See Netzhammer, "Self-Regulation in Broadcasting."
4. See Black, *The Catholic Crusade against the Movies*.
5. See Leff, *The Dame in the Kimono*. Leonard Leff reminds readers that in 1929 legislation was proposed that would have made the motion picture industry answerable to the Federal Trade Commission (FTC).
6. Holt, "*NYPD Blue*: Content Regulation," 271.
7. Hendershot, *Saturday Morning Censors*, 23.
8. Krattenmaker and Metzger, "FCC Regulatory Authority over Commercial Television Networks," 408.
9. Boddy, *Fifties Television*, 117.
10. Murray, "Television Wipes Its Feet," 133.
11. Meyer, "New Directions in Critical Television Studies," 265.
12. Wasko and Meehan, "Critical Crossroads or Parallel Routes?," 150.
13. Meehan, "Watching Television," 240.
14. Ibid., 240–241.
15. Meehan, "Critical Theorizing on Broadcast History," 31, 33.
16. Ibid., 41.
17. Mansfield, Welton, and Grogan, "'Truth or Consequences'," 1156.
18. Ibid., 1157.
19. Hawkesworth, "Policy Studies within a Feminist Frame," 99.
20. See Fischer, *Reframing Public Policy*. See also Diem et al., "The Intellectual Landscape of Critical Policy Analysis," 1069.
21. Diem et al., "The Intellectual Landscape of Critical Policy Analysis," 1069.
22. Luke, "What Is Critical?"
23. Fischer, "What Is Critical?," 95–96.
24. Ibid., 96.

25. Howarth, Glynos, and Griggs, "Discourse, Explanation, and Critique," 100.
26. Ibid.
27. See Streeter, *Selling the Air*; see also Holt, *Empires of Entertainment*.
28. NARTB, *The Television Code*, 1.
29. Ibid., 2–3.
30. Ibid., 3.
31. Ibid.
32. Ibid.
33. Ibid., 4.
34. Ibid., 8.
35. Ibid.
36. Ibid., 8–9.
37. Justin Miller to Frank Russell, August 28, 1951; Folder 8: J. Miller, May–August 1951; Box 93; National Association of Broadcasters Records (hereafter NAB Records); US MSS 156AF, Wisconsin Historical Society, Madison (hereafter WHS).
38. Clarke, *Regulation*, 2.
39. Baldwin, Cave, and Lodge, *Understanding Regulation*, 137.
40. Clarke, *Regulation*, 38.
41. The off-network syndication market is another, far more valuable source of money for the companies that own the rights to programs, but these rights holders reap the financial benefits of syndication *after* the programs have had their first run on a network.
42. Streeter, *Selling the Air*, 209.
43. Ibid., 210–213.
44. National Industrial Conference Board, *Trade Associations and Their Economic Significance and Legal Status*, 181.
45. Ibid., 182.
46. Ibid., 185.
47. Clarke, *Regulation*, 56.
48. Schatz, *Hollywood Genres*, 5–6; Neale, *Genre and Hollywood*, 23; Feuer, "Genre Study and Television," 142; Forman, "Television before Television Genre," 6–7.
49. Feuer, "Genre Study and Television," 142.
50. National Industrial Conference Board, *Trade Associations and Their Economic Significance and Legal Status*, 306.
51. Levin, "The Limits of Self-Regulation," 634.
52. National Industrial Conference Board, *Trade Associations and Their Economic Significance and Legal Status*, 192.
53. Ibid., 194.
54. Levin, "The Limits of Self-Regulation," 632, 604.
55. Ibid., 604.
56. Doherty, *Cold War, Cool Medium*, 61.
57. Murray, "'The Tendency to Deprave and Corrupt Morals,'" 141.
58. Besen et al., *Misregulating Television*, 25.
59. Peter J. Boyer, "Under Fowler, FCC Treated TV as Commerce," *New York Times*, January 19, 1987.
60. "Hollywood and TV," *Life*, February 25, 1952, 20.
61. Sewell, *Television in the Age of Radio*, 128.

Chapter 1

1. Douglas, *Inventing American Broadcasting*, 217.
2. FCC, "Regulation of Wire and Radio Communication," May 1, 1951; Folder 12–12: Organization, Federal Communications Commission, January 1, 1947–November 30, 1951; Box 8; General Correspondence, 1947–1956; Office of the Executive Director; FCC, Record Group 173 (hereafter RG 173); National Archives at College Park, MD (hereafter NACP).
3. Douglas, *Inventing American Broadcasting*, 220.
4. Ibid., 218.
5. Ibid., 236–237.
6. Tillinghast, *American Broadcasting Regulation and the First Amendment*, 18.
7. Ibid.
8. Ibid., 19.
9. Streeter, *Selling the Air*, 78.
10. Douglas, *Inventing American Broadcasting*, 237.
11. Arbuckle, "Herbert Hoover's National Radio Conferences," 13.
12. Aitken, "Allocating the Spectrum," 688.
13. Arbuckle, "Herbert Hoover's National Radio Conferences," 13.
14. Aitken, "Allocating the Spectrum," 693.
15. Benjamin, "Working It Out Together," 222.
16. Streeter, *Selling the Air*, 89.
17. McChesney, *Telecommunications, Mass Media, and Democracy*, 13.
18. Ibid., 14.
19. Ibid., 16.
20. Streeter, *Selling the Air*, 89.
21. Benjamin, "Working It Out Together," 230.
22. Sewell, *Television in the Age of Radio*, 94.
23. Ibid., 93–94.
24. Ibid., 94.
25. Scarpa, "The Anticompetitive Effects of Minimum Quality Standards," 45.
26. Arbuckle, "Herbert Hoover's National Radio Conferences," 56. Arbuckle provides a detailed analysis of each radio conference; see also Benjamin, "Working It Out Together," 221–236.
27. FCC, "Regulation of Wire and Radio Communication," May 1, 1951; Folder 12–12: Organization, Federal Communications Commission, January 1, 1947–November 30, 1951; Box 8; RG 173; NACP.
28. McChesney, *Telecommunications, Mass Media, and Democracy*, 17.
29. FCC, "Regulation of Wire and Radio Communication," May 1, 1951; Folder 12–12: Organization, Federal Communications Commission, January 1, 1947–November 30, 1951; Box 8; RG 173; NACP.
30. Perlman, *Public Interests*, 15.
31. McChesney, *Telecommunications, Mass Media, and Democracy*, 18.
32. Ibid., 22.
33. Ibid., 23.
34. Benjamin, "Working It Out Together," 233; McChesney, *Telecommunications, Mass Media, and Democracy*, 3.

35. McChesney, *Telecommunications, Mass Media, and Democracy*, 25.

36. Ibid., 27.

37. Ibid.

38. Ibid., 26.

39. Ibid., 29.

40. Holt, *Empires of Entertainment*, 54.

41. Benjamin, "Working It Out Together," 233.

42. Ibid.

43. McChesney, *Telecommunications, Mass Media, and Democracy*, 18.

44. Mackey, "The National Association of Broadcasters," 24–25.

45. Ibid., 27.

46. Harold E. Fellows, untitled speech, April 2, 1952; Folder 9: Speech–NARTB Convention, April 1952; Box 4; NAB Papers; WHS.

47. FCC, *Seventeenth Annual Report*, 13.

48. Ibid.

49. Barnouw, *A Tower in Babel*, 121.

50. National Industrial Conference Board, *Trade Associations and Their Economic Significance and Status*, 7.

51. Ibid., 15–16.

52. Ibid., 57.

53. Mackey acknowledges, however, that he wrote his dissertation with the cooperation of the NAB. He writes: "The officers and directors of the Association have, for the first time, opened their files to an outside researcher." Mackey, "The National Association of Broadcasters," 8.

54. Ibid., 2.

55. Streeter, *Selling the Air*, 5.

56. Mackey, "The National Association of Broadcasters," 3.

57. Ibid., 4.

58. Ibid., 7–8.

59. National Industrial Conference Board, *Trade Associations and Their Economic Significance and Legal Status*, 315–316.

60. Streeter, *Selling the Air*, 35.

61. Jassem, "An Examination of Self-Regulation of Broadcasting," 54.

62. Ibid., 51.

63. Tuchman, *The TV Establishment*, 15.

64. Ibid.

65. Harry Bannister to Milton L. Greenebaum, November 2, 1948; Folder 9: J. Miller, B–General Correspondence; Box 97; NAB Papers; WHS.

66. Ralph W. Hardy, untitled speech, June 23, 1950; Folder 14: Ralph Hardy–General, 1950–1954; Box 5A; NAB Papers; WHS.

67. Hendershot, *Saturday Morning Censors*, 14.

68. Ibid.

69. Tillinghast, *American Broadcasting Regulation and the First Amendment*, 53–54.

70. Radio Act of 1927, 47 U.S.C. §29 (1927).

71. Tillinghast, *American Broadcasting Regulation and the First Amendment*, 54.

72. Silverman, *You Can't Air That*, 3.

73. Jaramillo, "Astrological TV."

74. Ibid. Because the FRC and FCC did not establish a firm rule about astrology but held that it was not in the public interest, license renewal hearing outcomes were inconsistent. The high-profile, Los Angeles–based astrologer Carroll Righter believed that he had to appeal to the FCC when the Television Code retained the 1948 radio code's ban on astrology. And viewers, incensed at Righter's plight, launched a letter-writing campaign to the FCC.

75. Murray, "Broadcast Content Regulation and Cultural Limits," 13.

76. See Mackey, "The National Association of Broadcasters," 347–422. See also Morgan, *The Television Code of the National Association of Broadcasters*, 38–72.

77. Mackey, "The National Association of Broadcasters," 350.

78. Ibid., 351.

79. Morgan, *The Television Code of the National Association of Broadcasters*, 40.

80. Ibid., 41.

81. Ibid., 43.

82. Mackey, "The National Association of Broadcasters," 358.

83. Ibid., 362.

84. Morgan, *The Television Code of the National Association of Broadcasters*, 45.

85. Ibid., 46.

86. Mackey, "The National Association of Broadcasters," 374.

87. Ibid., 376.

88. Morgan, *The Television Code of the National Association of Broadcasters*, 47.

89. Mackey, "The National Association of Broadcasters," 361.

90. Ibid., 408.

91. Morgan, *The Television Code of the National Association of Broadcasters*, 52.

92. Pickard, "The Battle over the FCC Blue Book," 172.

93. FCC, *Public Service Responsibility of Broadcast Licensees*, 12.

94. Miller, *The Blue Book*, 3.

95. Ibid., 4.

96. Ibid., 8. Significantly, Miller claimed that "no broadcasters had any opportunity to participate in these determinations" of income as related to investment and revenues, but from the internal NAB memos discussed earlier in this section, we know that the FCC made a habit of communicating directly with stations regarding their income.

97. Ibid., 11.

98. Ibid., 12–13.

99. ACLU, *Radio Programs in the Public Interest*.

100. Ibid.

101. Ed Keys, "House Rushes Action on Probe of FCC," *Broadcasting*, June 21, 1948, 4; Ed Keys, "FCC Probe: Committee, Counsel Named," *Broadcasting*, July 5, 1948, 28; Ed Keys, "FCC Probe: Investigators Reported at Work," *Broadcasting*, July 26, 1948, 25.

102. Ed Keys, "Woodshed for FCC," *Broadcasting*, October 18, 1948, 21; Ed Keys, "FCC Probe: New Chairman May Sound Death Knell," *Broadcasting*, November 22, 1948, 25; "FCC Probe Off," *Broadcasting*, December 13, 1948, 23.

103. "Report on FCC," *Broadcasting*, January 10, 1949, 68.

104. "Coy Defends FCC against Censorship Charges," *Broadcasting*, January 24, 1949, 23.

105. Pickard, "The Battle over the FCC Blue Book," 186.

106. Morgan, *The Television Code of the National Association of Broadcasters*, 61.

107. NAB, *Standards of Practice for American Broadcasters*, 1.

108. Ibid., 2–3.

109. Ibid., 5.

110. See Morgan (*The Television Code of the National Association of Broadcasters*) for more on how the 1948 code functioned as a foundational document for the 1952 Television Code.

111. Murray, "Broadcast Content Regulation and Cultural Limits," 189.

112. Ibid., 190.

Chapter 2

1. This chapter is a modified version of Jaramillo, "The Rise and Fall of the Television Broadcasters Association, 1943–1951."

2. Sewell, *Television in the Age of Radio*, 144.

3. Wilbur, "The History of Television in Los Angeles, 1931–1952," 68–69.

4. Ibid., 69; "TBA Profile," *Television*, October 1946, 17.

5. Television Broadcasters Association (TBA), *What Is TBA Doing for the Television Industry?* (TBA publication), 1947; Folder 612: Television Broadcasters Association 1947; National Broadcasting Company History Files (hereafter NBC History Files); Library of Congress, Washington, DC (hereafter LOC).

6. "Television Broadcasters Assn. Formed by Engineers at Convention in Chicago," *Broadcasting*, January 24, 1944, 16.

7. Anderson, *Hollywood TV*, 37.

8. Hilmes, *Hollywood and Broadcasting*, 74.

9. O. B. Hanson to John H. MacDonald, January 30, 1947; Folder 612: Television Broadcasters Association 1947; NBC History Files; LOC.

10. Prior to his hiring by Paramount, Landsberg was employed by DuMont Laboratories.

11. TBA, *Proceedings of the Second Television Conference and Exhibition, Television Broadcasters' Association, Inc.*, 34.

12. Ibid., 36; "Television Broadcasters Association," *Television*, Spring 1944, 9.

13. "TBA Profile," *Television*, 17.

14. "Television Broadcasters Assn. Formed by Engineers at Convention in Chicago," *Broadcasting*, 16.

15. Ibid.

16. Ibid.

17. TBA, *Proceedings of Television Clinic of Television Broadcasters Association, Inc.*, 28.

18. TBA, *What Is TBA Doing for the Television Industry?*

19. TBA, "Annual Report to Members of Television Broadcasters Association," January 7, 1947; Folder 612: Television Broadcasters Association 1947; NBC History Files; LOC.

20. J. R. Poppele, "Annual Report of the President of Television Broadcasters Ass'n, Inc.," December 8, 1950; Folder 611: Television Broadcasters Association 1933–1934; NBC History Files; LOC (emphasis in original).

21. TBA, *A Pledge to Television Broadcasters* (TBA publication), 1950; Folder 611: Television Broadcasters Association 1933–1934; NBC History Files; LOC.

22. TBA, *What Is TBA Doing for the Television Industry?* (TBA publication), 1947; Folder 612: Television Broadcasters Association, 1947; NBC History Files, LOC.

23. Ibid.

24. For more on the TBA's involvement in FM allocations, see Jaramillo, "The Rise and Fall of the Television Broadcasters Association, 1943–1951," 6–7.

25. For a detailed explanation of the struggle over UHF and VHF television, see Boddy, *Fifties Television.*

26. TBA, *A Pledge to Television Broadcasters* (TBA publication), 1950; Folder 611: Television Broadcasters Association, 1933–1934; NBC History Files; LOC.

27. William A. Roberts, "Report to the Members of the Television Broadcasters Association," January 7, 1947; Folder 612: Television Broadcasters Association, 1947; NBC History Files; LOC.

28. TBA, "Annual Report to Members of Television Broadcasters Association, January 7, 1947; Folder 612: Television Broadcasters Association, 1947; NBC History Files, LOC. Such manufacturers included Farnsworth Television & Radio Corporation, Raytheon Manufacturing Company, Philco Corporation, Westinghouse Electric Corporation, and Belmont Electric Company. For a complete list of TBA members in 1947, see TBA, *What Is TBA Doing for the Television Industry?* (TBA publication), 1947; Folder 612: Television Broadcasters Association, 1947; NBC History Files, LOC.

29. TBA, "Annual Report to Members of Television Broadcasters Association," January 7, 1947; Folder 612: Television Broadcasters Association, 1947; NBC History Files, LOC.

30. Ernest A. Marx, DuMont Laboratories' television manager, first brought this to the TBA's attention in a special meeting in December 1946. See TBA, "Digest of a Special Meeting on December 19, 1946 of the Board of Directors of Television Broadcasters Association, Incorporated" (report), December 19, 1946; Folder 612: Television Broadcasters Association, 1947; NBC History Files; LOC.

31. TBA, "Digest of a Special Meeting on March 13, 1947 of the Board of Directors of Television Broadcasters Association, Incorporated" (report), March 13, 1947; Folder 612: Television Broadcasters Association, 1947; NBC History Files; LOC.

32. Spigel, *Make Room for TV*, 32.

33. Ibid. Between 1940 and 1950, New York City's population grew from 7,454,995 to 7,891,957. See US Bureau of the Census, *1950 Census of Population.*

34. TBA, *A Review of TBA Activities* (TBA publication), March 25, 1947; Folder 612: Television Broadcasters Association, 1947; NBC History Files; LOC.

35. Ernest A. Marx, "Annual Report on the Executive Committee on Affiliates to the Members of Television Broadcasters Association, Inc.," December 10, 1947; Folder 612: Television Broadcasters Association, 1947; NBC History Files; LOC.

36. J. R. Poppele, "Annual Report to Members of Television Broadcasters Association, Inc.," December 10, 1947; Folder 612: Television Broadcasters Association, 1947; NBC History Files; LOC.

37. J. R. Poppele to Joseph Nunan, March 20, 1947; Folder 612: Television Broadcasters Association, 1947; NBC History Files; LOC.

38. Joseph Nunan to J. R. Poppele, n/d; Folder 612: Television Broadcasters Association, 1947; NBC History Files; LOC.

39. J. R. Poppele, "Annual Report to Members of Television Broadcasters Association, Inc.," December 10, 1947; Folder 612: Television Broadcasters Association, 1947; NBC History Files, LOC.

40. TBA, "Minutes of a Special Meeting of the Board of Directors of Television Broadcasters Association, Inc.," November 6, 1947; Folder 612: Television Broadcasters Association, 1947; NBC History Files; LOC.

41. TBA, *A Pledge to Television Broadcasters* (TBA publication), 1950; Folder 611: Television Broadcasters Association, 1933–1934; NBC History Files; LOC.

42. Ibid.

43. J. R. Poppele, "Annual Report to Members of Television Broadcasters Association, Inc.," December 10, 1947; Folder 612: Television Broadcasters Association, 1947; NBC History Files; LOC.

44. Spigel, *Make Room for TV*, 25.

45. Ibid., 25–26.

46. Sewell, *Television in the Age of Radio*, 95.

47. Ibid., 115.

48. Will Baltin to Noran E. Kersta, August 9, 1948; Folder 615: TBA—Code Committee, 1948; NBC History Files; LOC. For the complete 1945 code, see TBA, *A Code to Guide the Presentation of Television Programs* (TBA publication), July 12, 1945; Folder 615: TBA—Code Committee, 1948; NBC History Files; LOC.

49. TBA, *A Pledge to Television Broadcasters* (TBA publication), 1950; Folder 611: Television Broadcasters Association, 1933–1934; NBC History Files; LOC.

50. J. R. Poppele, "Annual Report to Members of Television Broadcasters Association," January 7, 1947; Folder 612: Television Broadcasters Association, 1947; NBC History Files; LOC.

51. For a detailed discussion of the various radio codes and their reasons for failure, see Morgan, *The Television Code of the National Association of Broadcasters*.

52. Noran E. Kersta to Ken R. Dyke, George Frey, James Nelson, and H. M. Beville, September 16, 1948; Folder 615: TBA—Code Committee, 1948; NBC History Files; LOC.

53. Ibid.

54. TBA, "Code Committee of the TBA to Station Owners" (letter draft), October 8, 1948; Folder 615: TBA—Code Committee, 1948; NBC History Files; LOC.

55. J. R. Poppele, "Annual Report to Members of Television Broadcasters Association, Inc.," December 10, 1947; Folder 612: Television Broadcasters Association, 1947; NBC History Files; LOC.

56. L. W. Loman to Lawrence Phillips, July 30, 1948; Folder 615: TBA—Code Committee, 1948; NBC History Files; LOC.

57. TBA, *A Code to Guide the Presentation of Television Programs* (TBA publication), July 12, 1945; Folder 615: TBA—Code Committee 1948; NBC History Files; LOC.

58. Ibid.

59. TBA Code Committee, *A Statement of Principles and Policy* (TBA publication), November 4, 1948; Folder 621: TBA—Releases, 1948; NBC History Files; LOC.

60. Ibid.

61. Lawrence Lowman to Noran Kersta, July 21, 1948; Folder 615: TBA—Code Committee, 1948; NBC History Files; LOC.

62. David Sarnoff, "Television Today and Tomorrow," speech delivered May 18, 1931; File 21–132: David Sarnoff (1930–1935); Box 9; William S. Hedges Papers (hereafter Hedges Papers); Library of American Broadcasting, College Park, MD (hereafter LAB).

63. David Sarnoff, "Television Progress," speech delivered September 13, 1947; File 24–132: David Sarnoff (1945–1949); Box 9; Hedges Papers; LAB.

64. "Horizons Unlimited," *Broadcasting*, May 24, 1948, 23.

65. Ibid.

66. Ken Baker to Justin Miller, June 16, 1948; Folder 9: Justin Miller–Television Department, 1945–1953; Box 112; NAB Records; WHS.

67. Walter J. Damm to "Operating Television Stations," June 22, 1948; Folder 2: J. Miller [Board of Directors]–Television Advisory Committee (TAC), 1948; Box 112; NAB Records; WHS.

68. Sterling, O'Dell, and Keith, *The Concise Encyclopedia of American Radio*, 306; "NAB's FM-TV Aims," *Broadcasting*, September 12, 1949, 26.

69. Walter J. Damm to Justin Miller, July 7, 1948; Folder 2: J. Miller [Board of Directors]–TAC, 1948; Box 112; NAB Records; WHS.

70. Justin Miller to Walter J. Damm, July 13, 1948; Folder 2: J. Miller [Board of Directors]–TAC, 1948; Box 112; NAB Records; WHS.

71. Walter J. Damm to Justin Miller, July 16, 1948; Folder 2: J. Miller [Board of Directors]–TAC, 1948; Box 112; NAB Records; WHS.

72. NAB, press release, August 6, 1948, Folder 2: J. Miller [Board of Directors]–TAC, 1948, Box 112, NAB Records, WHS; A. D. Willard to Michael R. Hanna, July 30, 1948, Folder 2: J. Miller [Board of Directors]–TAC, 1948, Box 112, NAB Records, WHS.

73. "Report of Meeting, Television Broadcasters and NAB Representatives," August 11, 1948; Folder 2: J. Miller [Board of Directors]–TAC, 1948; Box 112; NAB Records; WHS.

74. NAB, press release, August 11, 1948; Folder 2: J. Miller [Board of Directors]–TAC, 1948; Box 112; NAB Records; WHS.

75. "Meeting . . . NAB Television Advisory Committee," memo drafted by A. D. Willard Jr., August 17, 1948; Folder 9: Justin Miller–Television Department, 1945–1953; Box 112; NAB Records; WHS. See also NAB Television Advisory Committee, meeting notes, August 13, 1948; Folder 2: J. Miller [Board of Directors]–TAC, 1948; Box 112; NAB Records; WHS.

76. NAB, press release, September 1, 1948; Folder 2: J. Miller [Board of Directors]–TAC, 1948; Box 112; NAB Records; WHS.

77. A. D. Willard Jr. to Justin Miller, September 15, 1948; Folder 9: Justin Miller–Television Department, 1945–1953; Box 112; NAB Records; WHS.

78. NAB, confidential memo, October 25, 1948; Folder 2: J. Miller [Board of Directors]–TAC, 1948; Box 112; NAB Records; WHS.

79. Will Baltin to Paul Raibourn, Noran E. Kersta, and G. Emerson Markham, October 26, 1948; Folder 622: TBA—TBA-NAB Merger, 1948; NBC History Files; LOC.

80. "NAB-BMB Facelifting," *Broadcasting*, November 22, 1948, 21.

81. Ibid., 74.

82. Ibid.

83. "TBA Election," *Broadcasting*, December 13, 1948, 34.

84. Ibid., 72.

85. "Coy Approached by TBA," *Broadcasting*, May 9, 1949, 23, 57.

86. "TBA Expansion," *Broadcasting*, February 7, 1949, 36.

87. "Statement to NAB Convention by J. R. Poppele," *Broadcasting*, April 11, 1949, 46 (emphasis in original).

88. "Coy Approached by the TBA," *Broadcasting*, May 9, 1949, 23.

89. Ibid.

90. See TBA, *A Pledge to Television Broadcasters* (TBA publication), 1950; Folder 611: Television Broadcasters Association, 1933–1934; NBC History Files, LOC. See also J. R. Poppele, "Annual Report to Members of Television Broadcasters Association, Inc.," December 10, 1947; Folder 612: Television Broadcasters Association, 1947; NBC History Files; LOC.

91. J. R. Poppele, "Annual Report to Members of Television Broadcasters Association, Inc.," December 10, 1947, Folder 612: Television Broadcasters Association, 1947, NBC History Files; LOC.

92. "BAB Policy Group," *Broadcasting*, May 9, 1949, 25.

93. Ibid.

94. "Blueprint for a Federated NAB," *Sponsor*, June 6, 1949, 28.

95. Ibid., 38.

96. Ibid., 36.

97. "Reorganize NAB?" *Broadcasting*, July 4, 1949, 23.

98. Justin Miller to J. M. McDonald, July 6, 1949; Folder 9: J. Miller, July 1949; Box 92; NAB Records; WHS.

99. "Streamlined NAB," *Broadcasting*, July 18, 1949, 23.

100. Ibid.

101. "NAB Revamping," *Broadcasting*, August 1, 1949, 26.

102. "NAB's FM-TV Aims," *Broadcasting*, September 12, 1949, 26.

103. Justin Miller to Clair McCollough, August 29, 1949; Folder 9: Justin Miller–Television Department, 1945–1953; Box 112; NAB Records; WHS. At the same time, Miller received word that ABC was looking to defect from the NAB.

104. "NAB's FM-TV Aims," *Broadcasting*, 40.

105. Ibid.

106. Ibid., 26.

107. Justin Miller to Hugh B. Terry, June 20, 1949; Folder 8: J. Miller, June 1949; Box 92; NAB Records; WHS.

108. Mark Woods to Justin Miller, July 21, 1949; Folder 7: Justin Miller, American Broadcasting Company, Inc., 1946–1953; Box 96; NAB Records; WHS.

109. Justin Miller to the NAB board of directors, February 24, 1950; Folder 14: J. Miller, January/February 1950; Box 92; NAB Records; WHS.

110. Eugene S. Thomas to Justin Miller, February 27, 1950, Folder 4: J. Miller–Networks, the NAB Membership and Relations, 1945–1950 (hereafter Folder 4), Box 108, NAB Records, WHS; John F. Meagher to Justin Miller, February 28, 1950, Folder 4, Box 108, NAB Records, WHS; Hugh B. Terry to Justin Miller, February 28, 1950, Folder 4, Box 108, NAB Records, WHS; Gilmore N. Nunn to Justin Miller, March 2, 1950, Folder 4, Box 108, NAB Records, WHS.

111. William B. Quarton to Justin Miller, February 28, 1950, Folder 4, Box 108, NAB Records, WHS; Robert D. Swezey to Justin Miller, March 3, 1950, Folder 4, Box 108, NAB Records, WHS; Charles C. Caley to Justin Miller, March 10, 1950, Folder 4, Box 108, NAB Records, WHS.

112. Harry R. Spence to Justin Miller, March 1, 1950; Folder 4; Box 108; NAB Records; WHS.

113. Justin Miller to William B. Quarton, May 4, 1950; Folder 4; Box 108; NAB Records; WHS.

114. "CBS Quits the NAB," *Broadcasting*, May 22, 1950, 23, 40.

115. Ibid., 23; Justin Miller to John J. Gillin Jr., June 7, 1950; Folder 2: J. Miller, April/May 1950; Box 93; NAB Records; WHS.

116. Justin Miller to Lester W. Lindow, July 4, 1950; Folder 2: J. Miller, April/May 1950; Box 93; NAB Records; WHS.

117. FCC, *Sixteenth Annual Report*, 5.

118. Hugh M. Beville Jr. to Meryl Sullivan, October 13, 1949; Folder 11: Justin Miller–Television Information Committee, 1950–1953; Box 112; NAB Records; WHS.

119. Justin Miller to Robert D. Swezey, July 25, 1950; Folder 11: Justin Miller–Television Information Committee, 1950–1953; Box 112; NAB Records; WHS.

120. "TBA to Expand," *Broadcasting*, June 5, 1950, 51.

121. Ibid.

122. Ibid.

123. J. R. Poppele, letter to the editor, *New York Times*, April 9, 1950, X11, ProQuest Historical Newspapers: *The New York Times* (1851–2009).

124. "Membership Figures of the NAB Summarized," *NAB Member Service* 2, no. 5 (1950).

125. "TBA to Expand," *Broadcasting*, 51.

126. "Television Committee Proposed Agenda for Meeting August 31 & September 1, 1950," n/d; Folder 11: Justin Miller–Television Information Committee, 1950–1953; Box 112; NAB Records; WHS.

127. Campbell Arnoux to Justin Miller, November 6, 1950; Folder 12: J. Miller–Television Planning and Promotion Committee, 1950–1951; Box 112; NAB Records; WHS.

128. Ibid.

129. Justin Miller to Robert D. Swezey, November 20, 1950; Folder 5: J. Miller, November/December 1950; Box 93; NAB Records; WHS.

130. "NAB Federation Takes Shape," *Broadcasting*, November 20, 1950, 28.

131. "Resolution Adopted Unanimously by the Board of Directors, National Association of Broadcasters," November 15, 1950; Folder 12: J. Miller–Television Planning and Promotion Committee, 1950–1951; Box 112; NAB Records; WHS.

132. Ibid.

133. NAB board of directors, telegram to 108 TV stations, November 16, 1950; Folder 12: J. Miller–Television Planning and Promotion Committee, 1950–1951; Box 112; NAB Records; WHS.

134. "Objectives for the NAB-TV Conference" (memo), January 11, 1951; Folder 12: J. Miller–Television Planning and Promotion Committee, 1950–1951; Box 112; NAB Records; WHS.

135. "Road Opens for NAB-TV," *Broadcasting*, December 4, 1950, 26.

136. Ibid.

137. Ibid.

138. Ibid.

139. "TBA-NAB to Confer on Merger Prospects," *Broadcasting*, December 11, 1950, 4.

140. Ibid.

141. "TV Gets Autonomous Trade Association," *NAB Member Service*, January 22, 1951.

142. "Video Industry Unity," *Broadcasting*, January 15, 1951, 51.

143. "Objectives for the NAB-TV Conference" (memo), January 11, 1951; Folder 12: J. Miller–Television Planning and Promotion Committee, 1950–1951; Box 112; NAB Records; WHS.

144. Ibid.

145. "TBA Out, NAB-TV In," *Television*, January 1951, 6.

146. Ibid.; Robert D. Swezey to Paul Raibourn, January 10, 1951, Folder 12, J. Miller–Television Planning and Promotion Committee, 1950–1951, Box 112, NAB Records, WHS.

147. NAB, form letter to TV stations, February 26, 1951, Folder 9: Justin Miller–Television Department, 1945–1953, Box 112, NAB Records, WHS; "TV into the NAB as Autonomous Entity," *Television Digest*, January 20, 1951, 2.

148. The NAB was a partner in the publication of *Broadcasting*. See McChesney, *Telecommunications, Mass Media, and Democracy*, 110; "NAB Gives Way to New Order . . . NARTB In?" *Broadcasting*, February 5, 1951, 24.

149. "NAB Becomes NARTB," *Broadcasting*, March 5, 1951, 23.

150. "TV into NAB as Autonomous Entity," *Television Digest*, January 20, 1951, 2.

151. "Autonomous NAB-TV Voted at Chicago Meet," *Broadcasting*, January 22, 1951, 4.

152. Ibid.

153. Justin Miller, J. R. Poppele, and Eugene S. Thomas to TV stations, n/d; Folder 9: Justin Miller–Television Department, 1945–1953; Box 112; NAB Records; WHS.

154. Ibid.

155. The NARTB changed its name back to NAB in 1958.

156. "NAB Becomes NARTB," *Broadcasting*, March 5, 1951, 23.

157. "NARTB as 'One Big Tent' with New Prexy," *Television Digest*, February 3, 1951, 3. See also "TV Gets Autonomous Trade Association," *NAB Member Service*, January 22, 1951.

158. Elected members of the TV board were Harold Hough, Clair McCullough, Robert D. Swezey, Paul Raibourn, George B. Storer, Harry Bannister, Campbell Arnoux, W. D. Rogers Jr., and Eugene Thomas.

159. "Name Change Proposed," *NAB Member Service*, January 22, 1951.

160. "NAB Becomes NARTB," *Broadcasting*, March 5, 1951, 23.

161. "NARTB as 'One Big Tent,'" *Television Digest*, February 3, 1951, 3. CBS returned in December 1951. See Martin Codel to Justin Miller, December 20, 1951; Folder 6: J. Miller, C–General Correspondence; Box 99; NAB Records; WHS.

162. "NARTB as 'One Big Tent,'" *Television Digest*, February 3, 1951, 3.

163. NBC, *Television Today: Its Impact on People and Products* (New York: NBC Television, 1951), 3; File 37–109: Television (A2–Z2); Box 8; Hedges Papers; LAB.

164. Ibid., 2.

165. Ibid., 5.

166. Ibid., 16–17, 20.

167. Ibid., 21.

168. Ibid.

Chapter 3

1. NAB, "What Is NAB?," n/d; Folder 1, NAB Publications: General, Promotional, and Membership Materials, 1940–ca. 1970; Box 118; NAB Records; WHS.

2. This chapter includes a version of Jaramillo, "Keep Big Government out of Your Television Set."

3. Aufderheide, *Communications Policy and the Public Interest*, 13.

4. Krasnow, Longley, and Terry, *The Politics of Broadcast Regulation*, 91.

5. TBA, "First Annual TBA Conference Draws Huge Crowd" (press release), December 11, 1944; Folder 14: Television Broadcasters Association; Box 9B; NAB Records; WHS.

6. Ibid.

7. Dr. Walter R. G. Baker, untitled address, December 11 and 12, 1944; Folder 14: Television Broadcasters Association; Box 9B; NAB Records; WHS.

8. Ibid.

9. TBA, "First Annual TBA Conference Draws Huge Crowd" (press release), December 11, 1944; Folder 14: Television Broadcasters Association; Box 9B; NAB Records; WHS.

10. TBA, *Proceedings of the First Annual Conference of the Television Broadcasters Association, Inc.*, 185.

11. Ibid., 1.

12. Ibid.

13. Ibid., 2.

14. Ibid., 9.

15. Ibid., 10.

16. Ibid., 88–89.

17. Ibid., 89.

18. Ibid., 75.

19. Ibid., 152.

20. Ibid.

21. Ibid., 15.

22. Ibid.

23. Ibid.

24. TBA, *Proceedings of the Second Television Conference and Exhibition, Television Broadcasters Association, Inc.*

25. Ibid., 41.

26. Ibid., "Foreword."

27. Ibid., 2–3.

28. Ibid., 3.

29. Ibid.

30. Ibid.

31. Ibid.

32. Ibid., 30.

33. Ibid., 31.

34. Ibid.

35. Ibid., 32.

36. TBA, *Proceedings of Television Clinic of Television Broadcasters Association, Inc.*, "Introduction."

37. Ibid., 31. According to the TBA's membership committee report, at the end of 1947 the TBA counted twenty-seven TV stations as members. Allen B. Dumont, "Membership Committee Report to Members of Television Broadcasters Association, Inc.," December 10, 1947; Folder 612: Television Broadcasters Association, 1947; NBC History Files; LOC. But in January 1948, Poppele stated before Congress that 80 percent

of "currently operating" TV stations were members. J. R. Poppele, President of the Television Broadcasters Association, statement before the House Committee on Education and Labor, January 14, 1948; Folder 621: TBA-Releases, 1948; NBC History Files; LOC.

38. TBA, *Proceedings of Television Clinic of Television Broadcasters Association, Inc.*, 35.

39. Ibid., 35.

40. Ibid., 36.

41. TBA, *A Code to Guide the Presentation of Television Programs* (TBA publication), July 12, 1945; Folder 615: TBA—Code Committee, 1948; NBC History Files; LOC.

42. Edward Lamb, "Anarchy in TV?: Broadcaster Says Video Needs Self-Control," *New York Times*, April 2, 1950, 105, ProQuest Historical Newspapers: *The New York Times* (1851–2009). For an extended analysis of Lamb's alleged Communist ties and his clashes with the FCC, see Brinson, *The Red Scare, Politics, and the Federal Communications Commission, 1941–1960*, 159–194.

43. J. R. Poppele, letter to the editor, *New York Times*, April 9, 1950, XII, ProQuest Historical Newspapers: *The New York Times* (1851–2009).

44. "Membership Figures of NAB Summarized," *NAB Member Service* 2, no. 5 (1950).

45. Poppele, letter to the editor, *New York Times*, April 9, 1950.

46. Clarke, *Regulation*, 41.

47. Ibid.

48. Doherty, *Cold War, Cool Medium*, 8.

49. Brinson, *The Red Scare, Politics, and the Federal Communications Commission, 1941–1960*, 3–4.

50. Ibid., 3.

51. Ibid., 3–4.

52. Ibid., 4.

53. Justin Miller, "Freedom of Communication," speech delivered May 23, 1947; Folder 6; Box 1; NAB Records; WHS.

54. Justin Miller, "Attacks on Freedom of Communication," speech delivered April 23, 1949; Folder 7: Speeches, Justin Miller, 1948–1959; Box 1; NAB Records; WHS.

55. Boddy, *Fifties Television*, 80–92.

56. Justin Miller to Walter J. Damm, Vice President and General Manager, WTMJ, March 25, 1950; Folder 1: NAB J. Miller, March 1950; Box 93; NAB Records; WHS.

57. Justin Miller to Kenneth Baker, March 1, 1950; Folder 1: J. Miller, March 1950; Box 93; NAB Records; WHS.

58. Justin Miller to *Christian Science Monitor*, June 9, 1950; Folder 2: NAB J. Miller, April–May 1950; Box 93; NAB Records; WHS.

59. Justin Miller to Wayne Coy, Chairman, FCC, May 1, 1950; Folder 2: NAB J. Miller, April–May 1950; Box 93; NAB Records; WHS.

60. Ralph Hardy, Director of Government Relations, untitled speech delivered June 23, 1950; Folder 14; Box 5A; NAB Records; WHS.

61. Justin Miller to A. D. Willard and Harold Fair, July 19, 1948; Folder 1: J. Miller, Interoffice Memoranda, May 1947–April 1949; Box 95; NAB Records; WHS.

62. "TV Code Plan," *Broadcasting*, November 8, 1948, 66.

63. Ibid.

64. NBC, *Responsibility: A Working Manual of NBC Program Policies* (New York: NBC, 1948); Folder 3: NBC Library: Program Policies, 1945–1956; Box 220; US MSS 17AF National Broadcasting Company Records (hereafter NBC Records); WHS. For a

comprehensive discussion of this manual, see Pondillo, *America's First Network TV Censor*, 55–69.

65. *NBC Standards and Practices Bulletin—No. 7: A Report on Television Program Editing and Policy Control*, November 1948; Folder 3: NBC Library: Program Policies, 1945–1956; Box 220; NBC Records; WHS.

66. Justin Miller to Ralph Hardy and G. Emerson Markham, February 23, 1950; Folder 3: J. Miller, Interoffice Memoranda, January–May 1950; Box 95; NAB Records; WHS.

67. "Coy on Television Programming" (reprint of press clipping), *Washington Evening Star*, February 21, 1950; Folder 20: Weaver, 1950: Policies; Box 119; NBC Records; WHS.

68. G. Emerson Markham to Sylvester L. Weaver, March 6, 1950; Folder 20: Weaver, 1950: Policies; Box 119; NBC Records; WHS.

69. Carleton D. Smith to G. Emerson Markham, March 13, 1950; Folder 20: Weaver, 1950: Policies; Box 119; NBC Records; WHS.

70. G. E. Markham to Justin Miller, March 27, 1950; Folder 5: J. Miller, Television Code, 1950–1951; Box 112; NAB Records; WHS.

71. "Adoption of Motion Picture Production Code for WOR-TV Announced by Theodore C. Streibert, President" (press release), March 27, 1950; Folder 5: J. Miller, Television Code, 1950–1951; Box 112; NAB Records; WHS.

72. Theodore C. Streibert to Justin Miller, March 28, 1950; Folder 5: J. Miller, Television Code, 1950–1951; Box 112; NAB Records WHS.

73. Justin Miller to Theodore C. Streibert, March 29, 1950; Folder 5: J. Miller, Television Code, 1950–1951; Box 112; NAB Records; WHS.

74. Walter J. Damm to Justin Miller, March 16, 1950; Folder 5: J. Miller, Television Code, 1950–1951; Box 112; NAB Records; WHS.

75. "Inviting a TV Code?" *Variety*, March 29, 1950; Folder 5: J. Miller, Television Code, 1950–1951; Box 112; NAB Records; WHS.

76. Ibid.

77. G. Emerson Markham to Justin Miller, April 7, 1950; Folder 5: J. Miller, Television Code, 1950–1951; Box 112; NAB Records; WHS.

78. "Inviting a TV Code?" *Variety*, March 29, 1950; Folder 5: J. Miller, Television Code, 1950–1951; Box 112; NAB Records; WHS.

79. G. E. Markham to Ralph Hardy, April 3, 1950; Folder 5: J. Miller, Television Code, 1950–1951; Box 112; NAB Records; WHS.

80. "American Television Society," in *The 1944 Radio Annual*, edited by Jack Alicoate (New York: The Radio Daily, 1944), 911.

81. Stockton Helffrich to Pat Weaver, April 25, 1950; Folder 56: Weaver, 1950: Continuity Acceptance; Box 118; NBC Records; WHS.

82. Helffrich questioned Breen on the disconnect between the Production Code's prohibition of "the flaunting of weapons" and the actual practice of gangster films. Breen replied that the Production Code was simply a public relations tool.

83. Pat Weaver to Stockton Helffrich, April 26, 1950; Folder 56: Weaver, 1950: Continuity Acceptance; Box 118; NBC Records; WHS.

84. G. Emerson Markham to Justin Miller, April 7, 1950; Folder 5: J. Miller, Television Code, 1950–1951; Box 112; NAB Records; WHS.

85. "Video Code Study: Lowman Heads TBA Unit," *Broadcasting*, May 8, 1950, 53.

86. Justin Miller to NAB board of directors, May 18, 1950; Folder 5: J. Miller, Television Code, 1950–1951; Box 112; NAB Records; WHS.

87. "TV Facing Need for a 'Morals Code,'" *Television Digest*, February 3, 1951, 3.

88. See NBC, *NBC Radio and Television Broadcast Standards* (New York: NBC, 1951); Folder 3: NBC Library: Program Policies, 1948–1956; Box 220; NBC Records; WHS.

89. Pondillo, *America's First TV Censor*, 69.

90. Fellows was elected president of the NARTB in April 1951, but Miller's and Fellows's new contracts did not go into effect until June of that year.

91. Press release, June 22, 1951; Folder 5: J. Miller, Television Code, 1950–1951; Box 112; NAB Records; WHS.

92. Press release, July 17, 1951; Folder 5: J. Miller, Television Code, 1950–1951; Box 112; NAB Records; WHS.

93. McCarthy, *The Citizen Machine*, 24 (emphasis in original).

94. Ibid.

95. For a detailed account of the drafting and adoption of the Code, see Morgan, *The Television Code of the National Association of Broadcasters*, 84–94.

96. Joseph McConnell to "All Television Production Personnel," November 7, 1951; Folder 18: Brooks: McConnell, Joseph, 1951; Box 130; NBC Records; WHS.

97. Harold E. Fellows, untitled speech delivered December 19, 1951; Folder 9: Harold E. Fellows Speeches; Box 1; NAB Records; WHS.

98. Harold E. Fellows, "Television—The Shape of Things to Come," March 21, 1952; Folder 10: Harold E. Fellows Speeches; Box 1; NAB Records; WHS.

99. MacCarthy, "Broadcast Self-Regulation," 673.

100. Harold E. Fellows, "Liberty—Let's Keep It," May 27, 1952; Folder 11: Harold E. Fellows Speeches; Box 1, NAB Records; WHS.

101. Harold E. Fellows, untitled speech delivered October 9, 1952; Folder 13: Harold E. Fellows Speeches; Box 1; NAB Records; WHS.

102. Harold E. Fellows, untitled speech delivered November 24, 1952; Folder 15: Harold E. Fellows Speeches; Box 1; NAB Records; WHS.

103. Justin Miller to Arthur L. Greene, Manager, KLTI and KLTI-FM, May 7, 1952; Folder 3: NAB J. Miller, March–June 1952; Box 94; NAB Records; WHS.

Chapter 4

1. T. J. Slowie to Miss Mary Boston, April 8, 1952; Folder 44–1; Box 41; RG 173; NACP. This quote is Slowie's phrasing of Boston's request.

2. Turow, "Another View of 'Citizen Feedback' to the Mass Media," 535.

3. Ibid.

4. Ibid., 540.

5. Simmons, "Dear Radio Broadcaster," 447.

6. Edith Obstfeld to FCC, October 24, 1952; Folder 44–3: Complaint File: Columbia Broadcasting System; Box 54; RG 173; NACP.

7. Turow, "Audience Construction and Culture Production," 104.

8. Ibid.

9. John F. Dille Jr., "Government Control of Broadcasting," *Chicago Today*, Spring 1966, 32.

10. Ibid.

11. Clarke, *Regulation*, 108.

12. Ibid.

13. Evelyn Anguiano to "President of the United States," November 19, 1952; Folder 44–1; Box 46; RG 173; NACP.

14. All of the following letters are from Folder 44–1; Box 46; RG 173; NACP: Rose Marie Kurds to FCC, n/d; Wayne Novak to FCC, December 23, 1952; Mrs. Mabel Schneider to FCC, December 29, 1952; Kathleen Griffin to FCC, December 29, 1952; Chuck Lassen to FCC, December 24, 1952; and Barbara Jean Hall to FCC, December 29, 1952.

15. Mrs. Mabel Schneider to FCC, December 29, 1952; Folder 44–1; Box 46; RG 173; NACP.

16. Apart from meriting an entire folder titled "Roller Derby," which unfortunately has been emptied, the following letters appear in the FCC papers: Forbes Shepherd to FCC, May 23, 1951, Folder 44–1, Box 41, RG 173, NACP; Nancy Pounds to FCC, August 17, 1951, Folder 44–3: American Broadcasting Company, Inc.—Program Complaints—Individual File, June 11, 1951–December 31, 1956 (hereafter ABC 1951–1956), Box 50, RG 173, NACP; Mildred G. Miller to unnamed recipient, n/d, Folder 44–3: ABC 1951–1956, Box 50, RG 173, NACP; Nancy Pounds to FCC, August 17, 1951, Folder 44–3: ABC 1951–1956, Box 50, RG 173, NACP; Mr. and Mrs. H. E. Ross to FCC, December 4, 1951, Folder 44–3: ABC 1951–1956, Box 50, RG 173, NACP.

17. T. J. Slowie to Mr. and Mrs. H. E. Ross, February 28, 1952; Folder 44–3; Box 50; RG 173; NACP.

18. Edwin M. Robinson to FCC, November 13, 1950; Folder 44–1; Box 40; RG 173; NACP.

19. Warren G. Magnuson to Wayne Coy, January 21, 1952; Folder 44–1; Box 45; RG 173; NACP.

20. William Boddy, *Fifties Television*, 81.

21. Feuer, "The Concept of Live Television," 16.

22. See Hilmes, *Hollywood and Broadcasting*, 143.

23. Sewell, *Television in the Age of Radio*, 120.

24. Hilmes, *Hollywood and Broadcasting*, 143.

25. George N. Beadstone to unnamed recipient, February 5, 1951; Folder 44–1; Box 41; RG 173; NACP.

26. Lawrence L. Cardwell to FCC, April 18, 1950; Folder 44–1; Box 40; RG 173; NACP.

27. Val Adams to ABC, CBS, and NBC, March 6, 1951; Folder 44–1; Box 45; RG 173; NACP.

28. A. W. Stubbs to FCC, n/d; Folder 44–1; Box 46; RG 173; NACP.

29. Roger G. Jajari to FCC, May 21, 1951; Folder 44–1; Box 45; RG 173; NACP.

30. Fred R. Kern to FCC, April 28, 1952; Folder 44–1; Box 45; RG 173; NACP.

31. Kompare, *Rerun Nation*, xv.

32. Ibid., 39.

33. N. M. Hunt to FCC, October 27, 1952; Folder 44–1; Box 46; RG 173; NACP.

34. Porst, *"United States v. Twentieth Century-Fox, et al.* and Hollywood's Feature Films on Early Television," 119.

35. Ibid.

36. Ibid., 120.

37. Ibid., 121–122.

38. Ibid., 122.

39. Ibid.

40. R. J. Winnager to FCC, August 26, 1949; File 44–1, Box 39; RG 173; NACP.

41. John Watson to the editor of the *Baltimore Evening and Morning Sun*, September 7, 1951; Folder 44–1; Box 45; RG 173; NACP.

42. Carl T. F. Newman to FCC, January 2, 1952, Folder 44–1, Box 45, RG 173, NACP; H. Munroe to FCC, May 10, 1952, Folder 44–1, Box 45, RG 173, NACP.

43. J. Fred Bellois to FCC, April 8, 1951; Folder 44–1; Box 45; RG 173; NACP.

44. Mrs. Robert Dowd to Wayne Coy, n/d; Folder 44–3: National Broadcasting Company, New York, New York, January 1, 1950–December 31, 1951 (hereafter NBC 1950–1951); Box 60; RG 173; NACP.

45. R. Kelly to FCC, June 6, 1952; Folder 44–3: National Broadcasting Company, New York, New York, January 1, 1952–June 30, 1955 (hereafter NBC 1952–1955); Box 61; RG 173; NACP.

46. A television set owner to Wayne Coy, January 14, 1950; Folder 44–1; Box 40; RG 173; NACP.

47. "Weekly Television Summary," *Broadcasting/Telecasting*, December 25, 1950, 64.

48. E. A. Hayl to FCC, October 1, 1950; Folder 44–1; Box 40; RG 173; NACP.

49. Lewis Epstein to FCC, August 27, 1951; Folder 44–1; Box 45; RG 173; NACP.

50. Rose Marie Macorama to FCC, February 13, 1951; Folder 44–3: Godfrey; Arthur, January 1, 1947–December 31, 1956 (hereafter Godfrey); Box 56; RG 173; NACP.

51. See H. R. Kimbrough to Paul A. Walker, October 8, 1952; Folder 44–3: Godfrey; Box 56; RG 173; NACP. For a discussion of Godfrey's style, popularity, and downfall, see Murray, "Our Man Godfrey."

52. J. Paul Buscher to FCC, December 27, 1951; Folder 44–3: NBC 1950–1951, Box 60; RG 173; NACP.

53. A. Carter to FCC, n/d; Folder 44–1; Box 45; RG 173; NACP.

54. Mrs. William S. Collyer to FCC, n/d; Folder 44–1; Box 41; RG 173; NACP.

55. T. J. Slowie to Mrs. William S. Collyer, July 25, 1952; Folder 44–1; Box 41; RG 173; NACP.

56. George Walton to FCC, June 7, 1951; Folder 44–1; Box 41; RG 173; NACP.

57. Harold Munroe to FCC, April 19, 1952, Folder 44–1, Box 41, RG 173, NACP; T. J. Slowie to Harold Munroe, May 22, 1952, Folder 44–1, Box 41, RG 173, NACP.

58. Donald Hall to FCC, September 23, 1952, Folder 44–1, Box 42, RG 173, NACP; T. W. Schreiber to FCC, August 31, 1951, Folder 44–1; Box 41, RG 173, NACP.

59. "Weekly Television Summary," *Broadcasting/Telecasting*, December 31, 1951, 56.

60. Robert M. Nelson to Television Department, ABC, CBS, NBC, and Dumont, March 14, 1950; Folder 44–1; Box 40; RG 173; NACP.

61. C. C. Cushman to FCC, n/d; Folder 44–1; Box 40; RG 173; NACP.

62. Leon W. Lopez to FCC, December 30, 1952; Folder 44–1; Box 46; RG 173; NACP.

63. Mrs. Harold M. Stern to Frank Samuels, November 1, 1950; Folder 44–3: American Broadcasting Company, Inc.–Program Complaints–Individual File, July 1, 1948–June 10, 1951 (hereafter ABC 1948–1951); Box 49; RG 173; NACP.

64. Salvatore Perri to FCC, December 14, 1951; Folder 44–3: NBC 1950–1951; Box 60; RG 173; NACP.

65. Robert Kuhn to David Sarnoff, January 1, 1951; Folder 44–3: NBC 1950–1951; Box 60; RG 173; NACP.

66. Henry L. Joynt to Wayne Coy, March 10, 1949; Folder 44–3: National Broadcasting Company, New York, New York, January 1, 1947–December 31, 1949 (hereafter NBC 1947–1949); Box 60; RG 173; NACP.

67. Juan A. Melendez to FCC, April 25, 1951; Folder 44–3: CBS 1948–1951; Box 65; RG 173; NACP.

68. Rev. Luiz G. F. Mendonça to FCC, n/d; Folder 44–3: NBC 1950–1951; Box 60; RG 173; NACP.

69. Clifford B. Reeves to Arthur Godfrey, April 17, 1950; Folder 44–3: Godfrey; Box 56; RG 173; NACP.

70. T. J. Slowie to Don Bartels, November 21, 1951; Folder 44–3: Complaint File–Columbia Broadcasting System, January 1, 1952–May 31, 1955 (hereafter CBS 1952–1955); Box 54; RG 173; NACP.

71. Wayne Coy to Sidney Correll, December 3, 1951; Folder 44–3: NBC 1950–1951; Box 60; RG 173; NACP.

72. T. J. Slowie to Donald F. Haynes, July 1, 1952; Folder 44–3: NBC 1952–1955; Box 61; RG 173; NACP.

73. Paul A. Walker to Thad Brown, August 14, 1952; Folder 44–1; Box 45; RG 173; NACP.

74. The following letters appear in Folder 44–1, Box 40, RG 173, NACP: W. H. O'Brien to FCC, January 14, 1950; William G. Greany to FCC, June 7, 1950; Lucy G. MacFarland to FCC, May 3, 1950; and R. M. Cheseldine to FCC, October 27, 1950. See also J. L. Schlotthauer, letter to FCC, December 6, 1952; Folder 44–1; Box 46; RG 173; NACP.

75. Wayne Coy to Franklin D. Roosevelt Jr., August 27, 1951; Folder 44–1; Box 45; RG 173; NACP.

76. Hendershot, *Saturday Morning Censors*, 1.

77. R. E. Roney to T. J. Slowie, January 19, 1952; Folder 44–1; Box 45; RG 173; NACP.

78. The Hubers to FCC, January 6, 1951; Folder 44–3: ABC 1948–1951; Box 49; RG 173; NACP.

79. Lucy G. MacFarland to FCC, May 3, 1950; Folder 44–1; Box 40; RG 173; NACP.

80. Irwin F. Bender to FCC, March 15, 1952, Folder 44–1, Box 45, RG 173, NACP; Kenneth L. Duncan to FCC, July, 31, 1949, Folder 44–1, Box 39, RG 173, NACP; Newton Rogers to FCC, November 7, 1949, Folder 44–1, Box 39, RG 173, NACP.

81. Etta C. Geis to FCC, July 5, 1951; Folder 44–1; Box 45; RG 173; NACP.

82. Francis X. Wallace to FCC, n/d; Folder 44–1; Box 45; RG 173; NACP.

83. Wendell P. C. Morgenthaler to FCC, April 1, 1952; Folder 44–1; Box 45; RG 173; NACP.

84. C. H. Newberry to Wayne Coy, February 6, 1952; Folder 44–1; Box 45; RG 173; NACP.

85. Tom Lyman to George E. Sterling, January 28, 1952; Folder 44–1; Box 45; RG 173; NACP.

86. "Irate Citizens," to FCC, February 1, 1951; Folder 44–1; Box 44; RG 173; NACP.

87. The precise numbers break down as follows: of the 353 letters focusing on indecency, 225 were devoted to costuming, 51 to jokes, 29 to sex, and 21 to burlesque. Multiple foci appeared in the same letters.

88. NARTB, *The Television Code*, 3.

89. John J. Conahan to FCC, April 5, 1949; Folder 44–1; Box 39; RG 173; NACP.

90. S. A. Wilder to FCC, February 27, 1951; Folder 44–1; Box 44; RG 173; NACP.

91. Harold W. Walker to FCC, March 21, 1951; Folder 44–3: CBS 1948–1951; Box 65, RG 173; NACP.

92. Joseph De Young to FCC, February 24, 1951; Folder 44–1; Box 44; RG 173; NACP.

93. William G. Mokray to FCC, January 25, 1951; Folder 44–3: Godfrey; Box 56; RG 173; NACP.

94. The Lofferts to FCC, n/d; Folder 44–1; Box 44; RG 173; NACP.

95. Bert McKasy to FCC, n/d; Folder 44–1; Box 45; RG 173; NACP.

96. Multiple authors to FCC, April 26, 1951; Folder 44–3: Godfrey; Box 56; RG 173; NACP.

97. Mrs. N. H. Millhiser to FCC, n/d; Folder 44–3: Godfrey; Box 56; RG 173; NACP.

98. Katherine Davids to FCC, February 13, 1951, Folder 44–1, Box 44, RG 173, NACP; Mrs. A. Kocher to FCC, December 24, 1951, Folder 44–1, Box 45, RG 173, NACP.

99. Susanne Kelly to FCC, February 6, 1951; Folder 44–1; Box 44; RG 173; NACP.

100. Schmidt to FCC, February 19, 1951; Folder 44–1; Box 44; RG 173; NACP.

101. Mrs. John Trogus to FCC, February 3, 1951; Folder 44–1; Box 44; RG 173; NACP.

102. Mrs. Evelyn Comstock to FCC, February 1, 1951; Folder 44–1; Box 44; RG 173; NACP.

103. "Just a Female Who is Totally Disgusted" to FCC, February 1, 1951; Folder 44–1; Box 44; RG 173; NACP.

104. Mrs. Roger K. Gingres to FCC, January 30, 1951; Folder 44–1; Box 44; RG 173; NACP.

105. Mrs. Marvin L. King to FCC, January 25, 1952; Folder 44–1; Box 45; RG 173; NACP.

106. Carol Wojcik to FCC, February 7, 1951; Folder 44–1; Box 44; RG 173; NACP.

107. Patricia Golden to FCC, February 8, 1951; Folder 44–1; Box 44; RG 173; NACP.

108. Mrs. O. H. Halverson to FCC, February 2, 1951; Folder 44–1; Box 44; RG 173; NACP.

109. M. Margaret Culleton to Wayne Coy, January 30, 1951; Folder 44–1; Box 44; RG 173; NACP.

110. Joseph A. Roberts to FCC, February 3, 1951; Folder 44–1; Box 44; RG 173; NACP.

111. Mrs. Joseph G. Bockelman to NBC, November 15, 1950; Folder 44–3: NBC 1950–1951; Box 60; RG 173; NACP.

112. Mrs. N. H. Millhiser to FCC, n/d, Folder 44–3: Godfrey, Box 56, RG 173, NACP; J. Reilly to FCC, January 30, 1951, Folder 44–3: Godfrey, Box 56, RG 173, NACP.

113. J. Reilly to FCC, January 30, 1951; Folder 44–3: Godfrey; Box 56; RG 173; NACP.

114. Mrs. Thomas A. Purtell to FCC, n/d; Folder 44–3: Wynn, Ed–Program Complaints–Individual File, March 1934–December 31, 1956; Box 65; RG 173; NACP.

115. Mrs. Wayne Campbell to FCC, n/d; Folder 44–3: Godfrey; Box 56; RG 173; NACP.

116. Mrs. Gladies Bishop to Wayne Coy, January 30, 1951; Folder 44–1; Box 44; RG 173; NACP.

117. Edgar A. Samuel to FCC, January 25, 1952; Folder 44–3: ABC 1951–1956; Box 50; RG 173; NACP.

118. Mrs. Harold M. Stern to Frank Samuels, November 1, 1950; Folder 44–3: ABC 1948–1951; Box 49; RG 173; NACP.

119. Dallas W. Smythe, *New York Television January 4–10, 1951, 1952: Monitoring Study No. 6* (Urbana, IL: National Association of Educational Broadcasters, 1952), 45, quoted in Barnouw, *The Golden Web*, 294.

120. Salvi S. Grupposo to FCC, March 7, 1952; Folder 44–1; Box 45; RG 173; NACP.

121. Alice Jones to FCC, February 6, 1952; Folder 44–1; Box 45; RG 173; NACP.

122. Wayne Coy, untitled speech delivered June 22, 1951; Folder 70–1; Box 84, RG 173; NACP.

123. NBC, *Responsibility*, 5.

124. See NBC, *NBC Radio and Television Broadcast Standards*, 4.

125. "Coy on Television Programming" (reprint of press clipping), *Washington Evening Star*, February 21, 1950; Folder 20: Weaver, 1950: Policies; Box 119; NBC Records; WHS.

126. Wayne Coy, untitled speech delivered June 22, 1951; Folder 70–1; Box 84; RG 173; NACP. Coy pointed out that while the 255 letters the FCC received about alcohol advertising mostly derived from an "organized campaign of the prohibition or temperance interests," the rest—including those relating to crime and indecency—were "spontaneous complaints."

127. NBC, *Responsibility*, 6.

128. Louis G. Norris to Senator Scott W. Lucas, June 8, 1949; Folder 44–1; Box 39; RG 173; NACP.

129. Schmidt to FCC, February 19, 1951; Folder 44–1; Box 44; RG 173; NACP.

130. G. C. Hunginser to FCC, March 22, 1949, Folder 44–1, Box 39, RG 173, NACP; Mrs. Augusta Levant to FCC, December 10, 1949, Folder 44–1, Box 39, RG 173, NACP.

131. Allene McCulloch to Wayne Coy, March 31, 1950; Folder 44–1; Box 39; RG 173; NACP.

132. Mrs. N. O. Gehrish to FCC, n/d; Folder 44–1; Box 45; RG 173; NACP; Mrs. John Derport to FCC, July 5, 1951; Folder 44–1; Box 45; RG 173; NACP; Miss Connelly to FCC, n/d; Folder 44–1; Box 41; RG 173; NACP.

133. Members of Spring Grove PTA to Representative James F. Lind, February 6, 1952; Folder 44–1; Box 45; RG 173; NACP.

134. Edith Krause to FCC, April 16, 1950; Folder 44–3: Colgate–Palmolive–Peet Co.–Program Complaints–Individual File, December 10, 1936–December 31, 1956; Box 53; RG 173; NACP.

135. NBC, *Responsibility*, 6.

136. Ibid.

137. See L. J. Bickelhaupt to FCC, December 21, 1949, Folder 44–1, Box 39, RG 173, NACP; Allene McCulloch to Wayne Coy, March 31, 1950, Folder 44–1, Box 39, RG 173, NACP.

138. Mrs. N. O. Gehrish to FCC, n/d; Folder 44–1; Box 45; RG 173; NACP. See also Martha Paulson to T. J. Slowie, n/d; Folder 44–3: ABC 1951–1956; Box 50; RG 173; NACP.

139. Mrs. Frank Bradstreet to FCC, May 5, 1951; Folder 44–1; Box 41; RG 173; NACP.

140. Martin, "You Don't Have to be a Bad Girl to Love Crime," 63.

141. Ibid.

142. See Alice Jones to FCC, February 6, 1952; Folder 44–1; Box 45; RG 173; NACP. See also Henry M. Saucen to FCC, June 22, 1951; Folder 44–1; Box 41; RG 173; NACP.

143. C. C. Cushman to FCC, n/d; Folder 44–1; Box 40; RG 173; NACP.

144. Mrs. E. Rombach to FCC, n/d; Folder 44–1; Box 45; RG 173; NACP.

145. Thomas R. Cox Jr. to Frank Stanton, October 24, 1951; Folder 44–3: CBS 1948–1951; Box 65; RG 173; NACP.

146. NBC, *Responsibility*, 6.

147. Hills, *The Pleasures of Horror*, 115.

148. Ibid.

149. See Waller, *American Horrors*, 148; Schmidt, "Television: Horror's 'Original' Home"; and Owens, "Coming Out of the Coffin."

150. See Killmeier, "More than Monsters."

151. Mrs. Catherine B. Hughes to FCC, n/d; Folder 44–1; Box 45; RG 173; NACP.

152. Janet Johnson to FCC, March 11, 1952; Folder 44–1; Box 45; RG 173; NACP.

153. Louis L. Weiner to FCC, May 31, 1952; Folder 44–1; Box 45; RG 173; NACP.

154. Paul A. Walker to James F. Lind, March 11, 1952; Folder 44–1; Box 45; RG 173; NACP.

155. Thomas R. Cox Jr. to Frank Stanton, October 24, 1951, Folder 44–3: CBS 1948–1951, Box 65, RG 173, NACP; Frank Kloch to FCC, February 23, 1952, Folder 44–3: NBC 1952–1955, Box 61, RG 173, NACP.

156. See Gordon Ellis to Frank Kloch, February 21, 1952; Folder 44–3: NBC 1952–1955; Box 61; RG 173; NACP.

157. Sidney Correll to Wayne Coy, December 11, 1951; Folder 44–3: NBC 1950–1951; Folder 60; RG 173; NACP.

158. McCusker, "'Dear Radio Friend,'" 176.

159. Ibid.

160. R. E. Roney to T. J. Slowie, December 1, 1951; Folder 44–1; Box 45; RG 173; NACP.

161. (Name unreadable) to T. J. Slowie, n/d; Folder 44–1; Box 45; RG 173; NACP.

162. R. E. Roney to T. J. Slowie, December 1, 1951; Folder 44–1; Box 45; RG 173; NACP.

163. T. J. Slowie to R. E. Roney, February 1, 1952; Folder 44–1; Box 45; RG 173; NACP.

164. E. N. Ricchezza to NBC, n/d; Folder 44–3: NBC 1950–1951; Box 60; RG 173; NACP.

165. Miriam M. Efron to T. J. Slowie, July 24, 1951; Folder 44–3: NBC 1950–1951; Box 60; RG 173; NACP.

166. L. Stauffer Oliver to FCC, June 1, 1949; Folder 44–1; Box 39; RG 173; NACP.

167. Henry Ward to FCC, January 30, 1950; Folder 44–1; Box 40; RG 173; NACP.

168. Robert M. Nelson to Television Department, ABC, CBS, NBC, and Dumont, March 14, 1950; Folder 44–1; Box 40; RG 173; NACP.

169. Sam Colacurcio to FCC, April 21, 1950; Folder 44–1; Box 40; RG 173; NACP.

170. Mrs. Alice R. Long to FCC, October 31, 1951; Folder 44–1; Box 41; RG 173; NACP.

171. Joseph Zingsheim to FCC, October 11, 1951; Folder 44–1; Box 41; RG 173; NACP.

172. H. H. Wilson to FCC, April 1, 1952; Folder 44–1; Box 41; RG 173; NACP.

173. Reverend Gardner L. Winn to FCC, April 11, 1950; Folder 44–1; Box 40; RG 173; NACP.

174. Henry M. Saucen to FCC, June 22, 1951; Folder 44–1; Box 41; RG 173; NACP.

175. Katherine F. Swanson to Wayne Coy, November 10, 1950; Folder 44–1; Box 40; RG 173; NACP.

176. Katherine Davids to FCC, February 13, 1951; Folder 44–1; Box 44; RG 173; NACP.

177. Mary Eichenlamb to FCC, February 8, 1951; Folder 44–1; Box 44; RG 173; NACP.

178. See Kathleen Riley to FCC, February 13, 1951; Folder 44–1; Box 44; RG 173; NACP. See also Marguerite Devlin to FCC, February 8, 1951; Folder 44–1; Box 44; RG 173; NACP.

179. Mrs. Lula M. Griffin to Wayne Coy, October 19, 1951; Folder 44–1; Box 45; RG 173; NACP.

180. Joseph F. Lamb to Wayne Coy, October 15, 1951; Folder 44–1; Box 45; RG 173; NACP.

181. Donald G. Livingston to FCC, January 30, 1951; Folder 44–1; Box 44; RG 173; NACP.

182. Peter Paul Kazenko to Wayne Coy, January 30, 1951; Folder 44–1; Box 44; RG 173; NACP.

183. Chick Houghton to FCC, September 26, 1952; Folder 44–1; Box 45; RG 173; NACP.

184. Agnes Hebard to FCC, June 6, 1952; Folder 44–1; Box 45; RG 173; NACP.

185. Clifford B. Reeves to Arthur Godfrey; April 17, 1950; Folder 44–3: Godfrey; Box 56; RG 173; NACP.

186. Edwin J. Beck to FCC, January 1952; Folder 44–3: ABC 1951–1956; Box 50; RG 173; NACP.

187. William J. Hill Jr. to WJZ-TV, January 1952; Folder 44–3: ABC 1951–1956; Box 50; RG 173; NACP.

188. David M. Martin to Celanese Corporation of America, February 9, 1952; Folder 44–3: ABC 1951–1956; Box 50; RG 173; NACP.

189. Howard E. Tower to Wayne Coy, December 5, 1950; Folder 44–1; Box 40; RG 173; NACP. For other letters requesting a code, see Mrs. Donald T. Nystrom to Wayne Coy, February 16, 1951, Folder 44–1, Box 44, RG 173, NACP; Reverend Victor M. Kolasa and Helen Rodzik, July 27, 1951, Folder 44–1, Box 45, RG 173, NACP; R. E. Roney to T. J. Slowie, December 1, 1951, Folder 44–3: ABC 1951–1956, Box 50, RG 173, NACP.

190. Isabel M. Grass to FCC, January 14, 1952; Folder 44–3: ABC 1951–1956; Box 50; RG 173; NACP.

191. T. J. Slowie to Mrs. Philip A. Brown, April 8, 1952; Folder 44–3: Godfrey; Box 56; RG 173; NACP.

192. Frank Klock to FCC, February 23, 1952; Folder 44–3: NBC 1952–1955; Box 61; RG 173; NACP.

193. Mrs. John Sarazen to FCC, May 18, 1952; Folder 44–1; Box 45; RG 173; NACP.

194. R. E. Roney to T. J. Slowie, April 21, 1952; Folder 44–1; Box 45; RG 173; NACP.

195. William Cape to FCC, September 3, 1952; Folder 44–1; Box 45; RG 173; NACP.

196. Arthur J. Freund to *St. Louis Post-Dispatch* editor, July 1, 1952; Folder 44–1; Box 45; RG 173; NACP.

197. Wallace Power to FCC, October 26, 1952; Folder 44–3: CBS 1952–1955; Box 54; RG 173; NACP.

198. Walter J. Damm to Pat Weaver, November 10, 1949; Folder 14: Weaver, 1949, Horror Stories on TV; Box 118; NBC Records; WHS.

199. George M. Burbach to Sylvester Weaver, December 7, 1949; Folder 14: Weaver, 1949, Horror Stories on TV; Box 118; NBC Records; WHS.

200. Carleton D. Smith to Sylvester L. Weaver Jr., November 21, 1949; Folder 14: Weaver, 1949, Horror Stories on TV; Box 118; NBC Records; WHS.

201. Syd Eiges to Pat Weaver, December 8, 1949; Folder 14: Weaver, 1949, Horror Stories on TV; Box 118; NBC Records; WHS.

202. Stockton Helffrich to Fred Wile, November 16, 1949; Folder 14: Weaver, 1949, Horror Stories on TV; Box 118; NBC Records; WHS.

203. Sylvester L. Weaver Jr. to Carleton Smith, George Frey, and Fred Wile, November 18, 1949; Folder 14: Weaver, 1949, Horror Stories on TV; Box 118; NBC Records; WHS.

204. Fred Wile, Jr. to Sylvester L. Weaver, Jr., memo, November 21, 1949; Folder 14: Weaver, 1949, Horror Stories on TV; Box 118; NBC Records; WHS.

205. Sam Kaufman to Syd Eiges, December 7, 1949; Folder 14: Weaver, 1949, Horror Stories on TV; Box 118; NBC Records; WHS.

206. Stockton Helffrich to Fred Wile Jr., December 9, 1949; Folder 14: Weaver, 1949, Horror Stories on TV; Box 118; NBC Records; WHS.

207. W. F. Brooks to Pat Weaver, November 28, 1950; Folder 53: Weaver, 1950: Complaints; Box 118; NBC Records; WHS.

208. Justin Miller to Edward Lamb, March 28, 1950; Folder 1: J. Miller, March 1950; Box 93; NAB Papers; WHS.

209. Ralph Hardy, testimony before a subcommittee of the House Committee on Interstate and Foreign Commerce, hearing on H. Res. 278, 82nd Cong. (1952), 141.

210. Wayne Coy, untitled speech delivered June 22, 1951; Folder 70–1; Box 84; RG 173; NACP.

211. William J. Hill, Jr. to WJZ-TV, January 1952; Folder 44–3: ABC 1951–1956; Box 50; RG 173; NACP.

212. Spigel, *Make Room for TV*, 36.

213. Ibid., 37.

214. Wayne Coy to Joseph F. Lamb, October 17, 1951; Folder 44–1; Box 45; RG 173; NACP.

215. Wang, "Convenient Fictions," 189.

216. Ibid.

217. Ibid., 190.

218. Ibid.

219. Paul A. Walker to Oren Harris, June 4, 1952; Folder 44–1; Box 45; RG 173; NACP.

Chapter 5

1. McChesney, *Telecommunications, Mass Media, and Democracy*, 23.

2. Streeter, *Selling the Air*, 15.

3. Ibid., 8, 21.

4. Landis, *Report on Regulatory Agencies to the President-Elect*, introduction (all subsequent citations refer to the Kindle edition); Krasnow, Longley, and Terry, *The Politics of Broadcast Regulation*, 9.

5. Krasnow, Longley, and Terry, *The Politics of Broadcast Regulation*, 9.

6. Freeman, "The Private Role in Public Governance."

7. Clarke, *Regulation*, 3, 4.

8. Ibid., 111.

9. Rabin, "Federal Regulation in Historical Perspective," 1193.

10. Ibid., 1196.

11. Horwitz, *The Irony of Regulatory Reform*, 46.

12. Rabin, "Federal Regulation in Historical Perspective," 1207.

13. Ibid., 1225.

14. Ibid., 1237.

15. Ibid., 1241–1242.

16. Ibid., 1262.

17. Ibid., 1263.

18. Freeman, "The Private Role in Public Governance," 545.

19. Robinson, "The Federal Communications Commission," 173.

20. Johnson, "A New Fidelity to the Regulatory Ideal," 875.

21. Robinson, "The Federal Communications Commission," 174.

22. Landis, *Report on Regulatory Agencies to the President-Elect*, introduction.

23. Ibid.

24. Baldwin, Cave, and Lodge, *Understanding Regulation*, 3.

25. Ibid.

26. Ibid., 15.

27. Prosser, "Regulation and Social Solidarity," 364.

28. Ibid., 365.

29. Ibid.

30. Ibid., 369.

31. Ibid., 372.

32. Ibid., 387.

33. Robinson, "The Federal Communications Commission," 174.

34. Krasnow, Longley, and Terry, *The Politics of Broadcast Regulation*, 15.

35. See Landis, *Report on Regulatory Agencies to the President-Elect*, introduction; see also Krasnow, Longley, and Terry, *The Politics of Broadcast Regulation*, 91.

36. Besen, Krattenmaker, Metzger, and Woodbury, *Misregulating Television*, 23.

37. Krasnow, Longley, and Terry, *The Politics of Broadcast Regulation*, 18.

38. Ibid.

39. Robinson, "The Federal Communications Commission," 176.

40. Freeman, "The Private Role in Public Governance," 546.

41. Noll, Peck, and McGowan, *Economic Aspects of Television Regulation*, 98.

42. Krasnow, Longley, and Terry, *The Politics of Broadcast Regulation*, 11.

43. Lowi, *The End of Liberalism*, 103.

44. Noll, Peck, and McGowan, *Economic Aspects of Television Regulation*, 98–99. See also Krasnow, Longley, and Terry, *The Politics of Broadcast Regulation*, 16.

45. The Supreme Court in *National Broadcasting Company v. United States* (1943) stated, among other things, that in denying licenses the FCC did not violate anyone's First Amendment protections. See Robinson, "The FCC and the First Amendment"; see also Goldberg and Couzens, "'Peculiar Characteristics.'"

46. John F. Dille Jr., "Government Control of Broadcasting," *Chicago Today*, Spring 1966, 30–35.

47. Ibid., 34.

48. Robinson, "The FCC and the First Amendment," 151. See also Krattenmaker and Powe, *Regulating Broadcast Programming*, 35–39.

49. Robinson, "The FCC and the First Amendment," 151.

50. Lowi, *The End of Liberalism*, 104.

51. Ibid.

52. Horwitz, *The Irony of Regulatory Reform*, 23.

53. Ibid.

54. Ibid., 26.

55. Ibid., 27.

56. Ibid.

57. Ibid., 28.

58. Ibid.

59. Ibid., 29.

60. Ibid., 31.

61. Ibid.

62. Ibid., 32.

63. Ibid., 39.

64. Ibid., 41.

65. Ibid., 43.

66. McChesney, *Telecommunications, Mass Media, and Democracy*, 18.

67. Johnson, "A New Fidelity to the Regulatory Ideal," 875.

68. Ibid., 883–884.

69. Krasnow, Longley, and Terry, *The Politics of Broadcast Regulation*, 139, 48.

70. Robinson, "The Federal Communications Commission," 190.

71. Ibid., 192.

72. Freeman, "The Private Role in Public Governance," 547.

73. Ibid., 548–549.

74. Landis, *Report on Regulatory Agencies to the President-Elect*, sect. II, subsect. C.

75. Ibid.

76. Clarke, *Regulation: The Social Control of Business between Law and Politics*, 117.

77. Ibid., 116.

78. Ibid., 117.

79. Ibid., 121.

80. Ibid.

81. Ibid.

82. Landis, *Report on Regulatory Agencies to the President-Elect*, sect. II, subsect. A, no. 5.

83. Mosco, *Broadcasting in the United States*, 26.

84. Streeter, *Selling the Air*, 148–151.

85. FCC Commissioner George E. Sterling, "Television—The One-Eye [*sic*] Monster and the FCC," speech delivered October 31, 1951; File 70–1: Speeches, Addresses, Etc. by Commission Staff & Commissioners; Box 84; RG 173; NACP (emphasis in original).

86. Waldo W. Primm to Representative Eugene Cox, September 19, 1951; File 12–12; Box 8; RG 173; NACP.

87. Noll, Peck, and McGowan, *Economic Aspects of Television Regulation*, 121.

88. Ibid.

89. Ibid., 122, 124.

90. Ibid., 121–122.

91. FCC Commissioner George E. Sterling, "Television—The One-Eye [*sic*] Monster and the FCC," speech delivered October 31, 1951; File 70–1, Box 84, RG 173; NACP.

92. FCC Commissioner Paul A. Walker, untitled speech delivered April 2, 1952; File 70–1; Box 84; RG 173; NACP.

93. FCC Commissioner E. M. Webster, "The Federal Communications Commission and Its Administrative Problems," speech delivered May 24, 1952; File 70–1; Box 84; RG 173; NACP. Commissioner Webster outlined in detail the problem with understaffing, pointing out that only thirteen staff members had been assigned to the Television Division. He wrote, "Think of it—thirteen persons consisting of six engineers, two lawyers,

two accountants, and three secretaries, to handle the expected avalanche of television applications. Obviously, they can't do it."

94. Cole and Oettinger, *Reluctant Regulators*, 8.

95. Robinson, "The Federal Communications Commission," 187.

96. Ibid.

97. Ibid., 188; Cole and Oettinger, *Reluctant Regulators*, 6.

98. Lichty, "Members of the Federal Radio Commission and Federal Communications Commission, 1927–1961," 28.

99. Ibid. Lichty's detailed breakdown is as follows: "Eleven commissioners have been from government service; eleven have had backgrounds in law; six have been in business, five have been journalists; five have been from the military, and two each have been educators, engineers or jurists."

100. Ibid.

101. Ibid., 30.

102. Ibid., 33–34.

103. Krasnow, Longley, and Terry, *The Politics of Broadcast Regulation*, 41.

104. Johnson, "A New Fidelity to the Regulatory Ideal," 885.

105. Noll, Peck, and McGowan, *Economic Aspects of Television Regulation*, 123; Cole and Oettinger, *Reluctant Regulators*, 8.

106. Noll, Peck, and McGowan, *Economic Aspects of Television Regulation*, 124.

107. Lichty, "Members of the Federal Radio Commission," 31.

108. Ibid., 31–32. Commissioners Henry A. Bellows and Sam Pickard both served as vice presidents of CBS, and FCC chair Charles R. Denny left the commission for a vice presidential post at NBC.

109. Mosco, *Broadcasting in the United States*, 27 (emphasis in original).

110. Freeman, "The Private Role in Public Governance," 571.

111. Ken Baker to Justin Miller, March 14, 1949; Folder 7: NAB J. Miller, FCC: General Correspondence, 1945–1953 (hereafter Folder 7: FCC); Box 102; NAB Records; WHS.

112. Memorandum for board of directors meeting, n/d; Folder 7: FCC; Box 102; NAB Records; WHS.

113. Robert T. Bartley to Justin Miller, October 19, 1945; Folder 7: FCC; Box 102; NAB Records; WHS.

114. Eugene Carr to Justin Miller, November 16, 1945; Folder 7: FCC; Box 102; NAB Records; WHS.

115. Justin Miller to NAB department heads, March 14, 1949; Folder 7: FCC; Box 102; NAB Records; WHS.

116. R. V. Howard to Justin Miller, April 5, 1949; Folder 7: FCC; Box 102; NAB Records; WHS.

117. Ken Baker to Justin Miller, March 14, 1949; Folder 7: FCC; Box 102; NAB Records; WHS.

118. Bob Richard to Justin Miller, March 14, 1949; Folder 7: FCC; Box 102; NAB Records; WHS.

119. Maurice B. Mitchell to Justin Miller, March 21, 1949; Folder 7: FCC; Box 102; NAB Records; WHS.

120. Harold Fair to Justin Miller, March 22, 1949; Folder 7: FCC; Box 102; NAB Records; WHS.

121. Don Petty to Justin Miller, April 5, 1951; Folder 7: FCC; Box 102; NAB Records; WHS.

122. A. D. Willard to Justin Miller, April 5, 1949; Folder 7: FCC; Box 102; NAB Records; WHS.

123. S. D. Suggs to Harold Fellows, January 26, 1952; William Benton Papers, (Box 360, Folder 8), Special Collections Research Center, University of Chicago Library (hereafter SCRC). A letter written to Harold Fellows stated, "You no doubt are aware of the reports scattered abroad that one Mr. Sarnoff of NBC in New York made an agreement with Mr. Denny, then the chairman of FCC, that if he would allow Mr. Sarnoff to write agreements between our FCC and similar organizations of other countries, that Mr. Sarnoff would in due time make Mr. Denny Vice President of NBC." The letter further asserted that Sarnoff had made the same agreement with Wayne Coy.

124. Justin Miller to Clair R. McCollough, January 6, 1951; Folder 6: NAB J. Miller, January–February 1951; Box 93; NAB Records; WHS.

125. Justin Miller, remarks, November 19, 1948; Folder 4: J. Miller, Freedom of Expression Conference Proceedings, 1948; Box 104; NAB Records; WHS.

126. FCC Commissioner Robert F. Jones, "Channels in the Sky," speech delivered January 17, 1950; File 70–1; Box 84; RG 173; NACP.

127. Justin Miller to Robert F. Jones, January 19, 1950; Folder 8: NAB J. Miller, General Correspondence (hereafter Folder 8); Box 106; NAB Records; WHS.

128. Robert F. Jones to Justin Miller, January 24, 1950; Folder 8; Box 106; NAB Records; WHS.

129. NAB, press release, January 27, 1950; Folder 8; Box 106; NAB Records; WHS. In a letter to Martin Codel of *Television Digest* the following year, Miller accused Jones of "getting emotionally upset and blowing his top." For more, see Justin Miller to Martin Codel, December 18, 1951; Folder 1: NAB J. Miller, November–December 1951; NAB Records; WHS.

130. Krasnow, Longley, and Terry, *The Politics of Broadcast Regulation*, 48.

131. Graham and Kramer, *Appointments to the Regulatory Agencies*.

132. Ibid., 18.

133. Ibid., 18–19.

134. Ibid., 18, 24.

135. Lichty, "Members of the Federal Radio Commission," 30.

136. Graham and Kramer, *Appointments to the Regulatory Agencies*, 25.

137. Ibid., 23.

138. Biographical sketch of Paul Atlee Walker, Chairman, Federal Communications Commission, March 5, 1952; Folder 20–14; Box 34; RG 173; NACP.

139. Biographical sketch of George E. Sterling, January 2, 1948, Folder 20–14, Box 34, RG 173, NACP; Zarkin and Zarkin, *The Federal Communications Commission*, 201.

140. Graham and Kramer, *Appointments to the Regulatory Agencies*, 22.

141. Flannery, *Commissioners of the FCC, 1927–1994*, 87–88.

142. Biographical sketch of Rosel H. Hyde, Chairman, Federal Communications Commission, April 20, 1953; Folder 20–14; Box 34; RG 173; NACP.

143. Graham and Kramer, *Appointments to the Regulatory Agencies*, 20.

144. FCC Chair Wayne Coy, untitled speech delivered April 18, 1950; Folder 70–1; Box 84; RG 173; NACP. Miller's title "The American Broadcaster's Responsibility to His Government" seems to have served as a topic assignment; the document itself has no title.

145. FCC Commissioner George Sterling, untitled speech delivered April 15, 1951; Folder 70–1; Box 84; RG 173; NACP.

146. Krasnow, Longley, and Terry, *The Politics of Broadcast Regulation*, 16.

147. Goldberg and Couzens, "'Peculiar Characteristics,'" 4.

148. See Rosenberg, "Program Content," 376, 379, 384; Boylan, "Legal and Illegal Limitations on Television Programming," 138–139; Loevinger, "The Issues in Program Regulation," 15; and Cox, "The FCC's Role in Television Programming Regulation," 593.

149. Robinson, "The FCC and the First Amendment," 111.

150. Rosenberg, "Program Content," 379.

151. Boylan, "Legal and Illegal Limitations on Television Programming," 140–141.

152. Fleming, "Television Programming," 17.

153. John F. Dille Jr., "Government Control of Broadcasting," *Chicago Today*, Spring 1966, 35.

154. Fleming, "Television Programming," 22.

155. Loevinger, "The Issues in Program Regulation," 8.

156. Robinson, "The FCC and the First Amendment," 119.

157. FCC Commissioner Frieda B. Hennock, "The Place of Radio and Television in the Future," speech delivered April 20, 1950, File 70–1, Box 84, RG 173, NACP; FCC Commissioner Frieda B. Hennock, "Seeing Is Believing," speech delivered November 15, 1950, File 70–1, Box 84, RG 173, NACP; FCC Commissioner Frieda B. Hennock, "TV—A Threat or a Blessing," speech delivered March 19, 1951, File 70–1, Box 84, RG 173, NACP; FCC Commissioner Frieda B. Hennock, "The Future of Television," speech delivered May 8, 1951, File 70–1, Box 84, RG 173, NACP. See also McChesney, *Telecommunications, Mass Media, and Democracy*, for a more nuanced understanding of how educators "lost" radio.

158. FCC Commissioner Frieda B. Hennock, "The Future of Television," speech delivered May 8, 1951; File 70–1; Box 84; RG 173; NACP.

159. FCC Chair Wayne Coy, untitled speech delivered May 5, 1950; File 70–1; Box 84; RG 173; NACP.

160. "FCC Won't Censor," *Broadcasting*, April 1, 1950, 28.

161. "Coy on Television Programming" (reprint of press clipping), *Washington Evening Star*, February 21, 1950; Folder 20: Weaver, 1950: Policies; Box 119; NBC Records; WHS.

162. "FCC Won't Censor," *Broadcasting*, April 1, 1950, 28.

163. "FCC Chairman Outlines Television Problem," *NAB Member Service*, March 20, 1950.

164. Robinson, "The FCC and the First Amendment," 121.

165. Cox, "The FCC's Role in Television Programming Regulation," 595.

166. Ibid., 596.

167. Loevinger, "The Issues in Program Regulation," 10.

168. "To Probe 'Anatomy' of TV Programs," *Television Digest*, January 27, 1951, 2–3.

169. Robinson, "The FCC and the First Amendment," 112.

170. Loevinger, "The Issues in Program Regulation," 7.

171. Goldberg and Couzens, "'Peculiar Characteristics,'" 15.

172. "TV 'Blue Book' Plans Official," *Broadcasting*, February 5, 1951, 21.

173. FCC Chairman Wayne Coy, untitled speech delivered June 22, 1951; File 70–1; Box 84; RG 173; NACP.

174. Ibid.

175. FCC Commissioner Frieda B. Hennock, "The Future of Television," speech delivered May 8, 1951; File 70–1; Box 84; RG 173; NACP.

176. FCC Commissioner E. M. Webster, "Community Interest in Television Programming," speech delivered August 13, 1951; File 70–1; Box 84; RG 173; NACP.

177. FCC Commissioner Paul A. Walker, "Broadcasting and the Church," speech delivered October 2, 1951; File 70–1; Box 84; RG 173; NACP.

178. "Commissioner Walker Sounds a Warning on Program Excesses by Telecasters and Broadcasters," *Television Digest*, October 6, 1951, "Special Report" section.

179. Ibid.

180. Justin Miller to Paul A. Walker, October 22, 1951; Folder 9: NAB J. Miller, September–October 1951; Box 93; NAB Records; WHS.

181. NARTB, press release, June 22, 1951; Folder 3: J. Miller, Television Code, 1950–1951; Box 112; NAB Records; WHS.

182. Frank M. Russell to Robert D. Swezey, June 28, 1951; Folder 3: J. Miller, Television Code, 1950–1951; Box 112; NAB Records; WHS.

183. Robert D. Swezey to Frank M. Russell, July 6, 1951; Folder 3: J. Miller, Television Code, 1950–1951; Box 112; NAB Records; WHS.

184. NARTB Television Program Standards Committee, "Proposed Minutes," July 30, 1951; Folder 3: J. Miller, Television Code, 1950–1951; Box 112; NAB Records; WHS.

185. FCC Commissioner George E. Sterling, "Television—The One-Eye [*sic*] Monster and the FCC," speech delivered October 31, 1951; File 70–1; Box 84; RG 173; NACP.

186. FCC Commissioner Paul A. Walker, "Broadcasting and Human Relations," speech delivered February 20, 1952; File 70–1; Box 84; RG 173; NACP.

187. Paul A. Walker, "FCC Outlook," *Broadcasting*, December 31, 1951, 60.

188. FCC Chair Paul A. Walker, "The Role of Federal Regulation of Broadcasting in American Democracy," speech delivered April 28, 1953; File 70–1; Box 84; RG 173; NACP.

189. Brenner, "The Limits of Broadcast Self-Regulation under the First Amendment," 1552.

190. Ibid.

191. Ibid., 1550.

192. Ibid., 1552–1553.

193. Freeman, "The Private Role in Public Governance," 648.

194. Netzhammer, "Self-Regulation in Broadcasting," v.

Chapter 6

1. Wayne Coy to Frederick J. Lawton, June 7, 1951; Folder 76–1; Box 85; RG 173; NACP.

2. Statement of Representative Thomas Lane, 97 Cong. Rec. H1701 (daily ed., February 28, 1951).

3. Ibid. Lane stated, "Once we could protect our children by example and training, developing in them the character and integrity to meet the problems of life gradually. We made sure that they did not associate with bad companions in their formative years. We

even supervised their entertainment. We could do this in advance, for example, because we could find out what movies were not good for them. At the present time, however—and right in the home—there is no way of knowing from moment to moment just what embarrassing or shameful surprise is in store for us and our children. This places parents in a position of perpetual uneasiness and helplessness."

4. "State and Local Censorship of Films Used on Television," *Federal Communications Bar Journal* 4 (Winter 1949): 194.

5. Ibid.

6. Ibid., 197.

7. Harper and Etherington, "What the Supreme Court Did Not Do during the 1950 Term," 373. Of course the Pennsylvania Board of Censors appealed, but the Third Circuit Court of Appeals upheld the original ruling. In 1951, the Supreme Court declined to review the ruling, so the Third Circuit's decision stood.

8. Wayne Coy, untitled speech delivered June 22, 1951; Folder 70–1; Box 84; RG 173; NACP.

9. Meyers, "From Radio Adman to Radio Reformer," 19.

10. Ibid., 21.

11. Ibid.

12. Ibid.

13. Ibid., 22.

14. Ibid., 23.

15. Ibid., 24.

16. William Benton, "Television with a Conscience," *Saturday Review of Literature*, August 25, 1951, 8, www.unz.org/Pub/SaturdayRev-1951aug25-00007.

17. Meyers, "From Radio Adman to Radio Reformer," 24–25.

18. Ibid., 25.

19. Sewell, *Television in the Age of Radio*, 98.

20. Ibid.

21. Benton, "Television with a Conscience," 7.

22. Ibid., 8.

23. Ibid.

24. Ibid., 8, 30.

25. Ibid., 30.

26. Ibid., 31.

27. Ibid.

28. For an exhaustive account of this opposition, see McChesney, *Telecommunications, Mass Media, and Democracy*.

29. Kenneth Baker, statement to FCC, January 26, 1951; Folder 10: J. Miller, FCC Hearings, Educational Television, Outline of Testimony, January 26, 1951; Box 102; NAB Records; WHS.

30. Perlman, *Public Interests*, 15.

31. Ibid., 18.

32. Benton, "Television with a Conscience," 31.

33. 97 Cong. Rec. S3824 (daily ed., April 13, 1951).

34. "Education Channels: Benton Asks Review," *Broadcasting*, April 16, 1951, 4.

35. 97 Cong. Rec. S3824 (daily ed., April 13, 1951).

36. Office of Senator William Benton, press release, April 19, 1951; William Benton Papers; Box 347; Folder 4; SCRC.

37. William Benton to Frank Stanton, April 22, 1952; William Benton Papers; Box 112; Folder 2; SCRC.

38. William Benton to Frank Stanton, April 26, 1951; William Benton Papers; Box 112; Folder 2; SCRC.

39. John Howe to William Benton, April 5, 1951; William Benton Papers; Box 360; Folder 6; SCRC.

40. Ibid. (emphasis in original).

41. Justin Miller to William Benton, Folder 8: J. Miller, May–August 1951; Box 93; NAB Records; WHS.

42. "Benton Blow Threatens," *NARTB Member Service: Confidential TV Newsletter*, May 1, 1951, 1–2.

43. Ibid., 2.

44. Ibid., 2–3.

45. Statement of Senator Benton, 97 Cong. Rec. S5301 (daily ed., May 15, 1951).

46. Ibid., 5302.

47. Ibid.

48. Ibid.

49. Ibid., 5303.

50. John Howe to William Benton, May 15, 1951; William Benton Papers; Box 347; Folder 4; SCRC.

51. Senator William Benton, "Use of Television Frequencies for Educational Purposes," statement before a subcommittee of the Senate Committee on Interstate and Foreign Commerce, hearing on S. Res. 127, 82nd Cong. (1951), 2.

52. William Benton to Paul Hoffman, May 21, 1951; William Benton Papers; Box 347; Folder 4; SCRC.

53. Benton, "Use of Television Frequencies for Educational Purposes," hearing on S. Res. 127, 82nd Cong. (1951), 16.

54. Ibid., 7.

55. Ibid., 19.

56. Ibid., 21.

57. "Freeze Lift Delay Urged by Benton Plan," *Broadcasting*, June 11, 1951, 25.

58. Ibid.

59. Perlman, *Public Interests*, 26.

60. Ibid.

61. Benton, "Television with a Conscience," 32.

62. Ibid.

63. Benton, "Use of Television Frequencies for Educational Purposes," hearing on S. Res. 127, 82nd Cong. (1951), 10.

64. Benton, "Television with a Conscience," 32.

65. "Freeze Lift Delay Urged by Benton Plan," *Broadcasting*, June 11, 1951, 25.

66. Ibid.

67. Mary Graves to William Benton, September 5, 1951; William Benton Papers; Box 360; Folder 8; SCRC.

68. Peter Perry to William Benton, September 8, 1951; William Benton Papers; Box 360; Folder 8; SCRC.

69. William Elliot to William Benton, September 14, 1951, William Benton Papers, Box 360, Folder 8, SCRC; Morris Goldstone to William Benton, May 11, 1951, William Benton Papers, Box 360, Folder 6, SCRC.

70. Beth K. Spencer to William Benton, October 23, 1951, William Benton Papers, Box 360, Folder 5, SCRC; William L. Gordon Jr. to William Benton, October 24, 1951, William Benton Papers, Box 360, Folder 5, SCRC.

71. Gordon L. Nelson to William Benton, October 22, 1951; Mrs. John Glaettli to William Benton, January 15, 1952; Cora Thorpe to William Benton, February 1, 1952; all in William Benton Papers, Box 360, Folder 5, SCRC.

72. "Freeze Lift View," *Broadcasting*, June 18, 1951, 60.

73. Ibid.

74. Ibid.

75. Eugene Carr to John Bricker, June 18, 1951; Folder 6: J. Miller, C–General Correspondence; Box 99; NAB Records; WHS.

76. See William Boddy, *Fifties Television*, 42–57.

77. Justin Miller to Harold E. Fellows, June 22, 1951; Folder 5: J. Miller Interoffice Memoranda, 1951–1953; Box 95; NAB Records; WHS.

78. Ibid.

79. Edwin H. James, "Educational Dilemma," *Broadcasting*, July 16, 1951, 58. See also Benton, "Use of Television Frequencies for Educational Purposes," hearing on S. Res. 127, 82nd Cong. (1951), 20–21.

80. Hearing of the Senate Committee on Interstate and Foreign Commerce on "FCC Policy on Television Freeze and Other Communication Matters," 82nd Cong. (1951).

81. Ibid., statement of Honorable Wayne Coy, Chairman, FCC, 6.

82. Ibid.

83. Ibid., 42.

84. Ibid., 43.

85. Ibid.

86. The memo was entered into the *Congressional Record* after the conclusion of the hearing; see hearing on "FCC Policy on Television Freeze and Other Communication Matters," 82nd Cong. (1951), 44–47.

87. Ibid., 48–49.

88. Ibid., 47.

89. Ibid., 49. Toward the end of the hearing, Senator Homer Capehart (R-IN) asked Coy for the definition of an educational program, to which Coy replied, "one that is put on by an educational institution and has to do with the improvement of the cultural background and understanding of the community." Ibid., 51. Coy further explained that public service programming and educational programming were two different things. Ibid., 52.

90. William Benton to Robert F. Jones, July 27, 1951; William Benton to George Edward Sterling, July 27, 1951; William Benton to Edward M. Webster, July 27, 1951; all in William Benton Papers, Box 360, Folder 4, SCRC.

91. William Benton to Robert Kintner, July 27, 1951; William Benton to Joseph McConnell, July 27, 1951; William Benton to Edward Noble, July 27, 1951; William Benton to William Paley, July 27, 1951; William Benton to Niles Trammel, July 27, 1951; all in William Benton Papers, Box 360, Folder 4, SCRC.

92. "Bill Benton's Bill," *NARTB Member Service: Confidential TV Newsletter*, August 1, 1951, 3.

93. Ibid.

94. Senator William Benton, statement before a subcommittee of the Committee on Interstate and Foreign Commerce, hearing on S. 1579 and S. J. Res. 76 to establish a National Citizens Advisory Board on Radio and Television, 82nd Cong. (1951), 8.

95. Ibid., 7.

96. Ibid., 8–9.

97. Ibid., 24.

98. Ibid.

99. Ibid.

100. Ibid., 26.

101. Ibid., 9.

102. Ibid., 10.

103. Ibid., 9.

104. "Advisory Board: Benton Offers Substitute," *Broadcasting*, August 20, 1951, 25.

105. Ibid.

106. William Benton, statement before a hearing on S. 1579 and S. J. Res. 76, 82nd Cong. (1951), 51.

107. Ibid., 32.

108. Ibid.

109. Ibid., 34.

110. Wayne Coy, Chairman, FCC, statement before hearing on "FCC Policy on Television Freeze and Other Communication Matters," 82nd Cong. (1951), 6. This point was raised in the freeze hearing, during which Benton grilled Coy about the FCC's reluctance to punish broadcasters for violating the terms of their licenses. Coy defended broadcasters' rights to have hearings and to appeal, citing a variation on "innocent until proven guilty."

111. Ibid., 35.

112. Ibid., 36.

113. Ibid.

114. Clarke, *Regulation*, 120.

115. William Benton to Robert Kintner, September 1, 1952; William Benton Papers; Box 102; Folder 3; SCRC.

116. "Testimony on Benton Measures Draws Immediate Fire from NARTB Television Board," *NARTB Member Service: Confidential Management Newsletter*, September 10, 1951, 1.

117. "More on Benton Measure," *NARTB Member Service: Confidential TV Newsletter*, October 1, 1951, 4.

118. "Bill Benton Rides Again," *NARTB Member Service: Confidential TV Newsletter*, September 1, 1951, 1.

119. "NARTB Board Meet," *Broadcasting*, September 10, 1951, 77.

120. Paul Porter to Harold Fellows, September 7, 1951; William Benton Papers; Box 360; Folder 9; SCRC.

121. William Benton to Justin Miller, September 10, 1951; William Benton Papers; Box 105; Folder 5; SCRC.

122. Justin Miller to William Benton, October 5, 1951; Folder 9: J. Miller, September–October 1951; Box 93; NAB Records; WHS.

123. Ibid. (emphasis in original).

124. William Benton to John W. Bricker, September 21, 1951; William Benton Papers; Box 360; Folder 4; SCRC.

125. Clarke, *Regulation*, 111.

126. Ibid.

127. Ibid.

128. "The Benton Bills—What Are They?" *NARTB Member Service: Government Relations*, September 19, 1951, 4.

129. "NARTB Board Meet," *Broadcasting*, September 10, 1951, 77.

130. Ibid.

131. Ibid.

132. Ibid. (emphasis added).

133. "Benton Plan: WLBH's Livesay Attacks," *Broadcasting*, September 24, 1951, 88.

134. Frank Stanton to William Benton, July 30, 1951; William Benton Papers; Box 112; Folder 2; SCRC.

135. "FCC Disapproves Benton Measures," *NARTB Member Service: Industry-Government Highlights*, October 15, 1951, 32.

136. "FCC Majority Blasts Benton Video Bill," *Broadcasting*, October 15, 1951, 5.

137. "FCC Fears 'Censorship' in Benton Bills," *Television Digest*, October 13, 1951, 4.

138. Ibid.

139. "Benton Undeterred on Board Plan," *Broadcasting*, October 22, 1951, 33.

140. Ibid.

141. "Statement by the American Civil Liberties Union Supporting in General, S. 1579, a Bill Establishing a National Citizens Advisory Board on Radio and Television," October 10, 1951; William Benton Papers; Box 360; Folder 9; SCRC.

142. William Benton to Patrick Murphy Malin, October 17, 1951; William Benton Papers; Box 360; Folder 4; SCRC.

143. "Board Meetings Open Today," *NARTB Member Service: Confidential Management Newsletter*, June 4, 1951, 1.

144. "TV Program Logs to Be Studied," *NARTB Member Service: Confidential Management Newsletter*, June 11, 1951, 3.

145. "TV Program Standards to Be Reviewed," *NARTB Member Service: Confidential Management Newsletter*, June 18, 1951, 3.

146. "Telecasters to Develop Standards," *NARTB Member Service: Confidential Management Newsletter*, June 25, 1951, 1.

147. "Senator Johnson Foresees No Censorship," *NARTB Member Service: Confidential Management Newsletter*, June 25, 1951, 2.

148. Ibid.

149. "Coy Addresses Telecasters," *NARTB Member Service: Confidential Management Newsletter*, June 25, 1951, 2.

150. "Telecasters to Develop Standards," *NARTB Member Service: Confidential Management Newsletter*, June 25, 1951.

151. "Congress Outlook: Leaders See Radio and TV in Front Ranks," *Broadcasting*, December 31, 1951, 28.

152. Charles R. Denny to Joseph H. McConnell, July 3, 1951; Folder 8; Box 130; NBC Records; WHS.

153. Boddy, *Fifties Television*, 160.

154. Ibid.

155. Ibid., 175.

156. Jack Gould, "Radio and Television: New TV Code of Ethics Held Effort to Counter Bill Creating Citizens Board to Review Programs," *New York Times*, October 29, 1951, 32, ProQuest Historical Newspapers: *The New York Times* (1851–2009).

157. Ibid.

158. Ibid.

159. Norman R. Glenn, "Anything-Goes' TV Era on Way Out," *Sponsor*, November 5, 1951, 83.

160. William Benton to Wayne Coy, January 3, 1952; William Benton Papers; Box 360; Folder 5; SCRC.

161. Mark M. MacCarthy, "Broadcast Self-Regulation: The NAB Codes, Family Viewing Hour, and Television Violence," *Cardozo Arts and Entertainment Law Journal* 13, no. 3 (1995): 675.

162. Justin Miller to Bob Richards, September 26, 1952; Folder 4: J. Miller, July–October 1952; Box 94; NAB Records; WHS.

163. William Benton to Maurice B. Mitchell, January 3, 1952; William Benton Papers; Box 360; Folder 5; SCRC.

164. "Telecasters Adopt Self-Control Code," *Television Digest*, October 20, 1951, 6.

165. "Benton Measures Continue to Be Vital Issue," *NARTB Member Service: Confidential Radio Newsletter*, October 29, 1951, 1.

166. "More about the Benton Bill," *NARTB Member Service: Confidential Radio Newsletter*, November 12, 1951, 3–4.

167. Harold E. Fellows, untitled speech delivered December 19, 1951; Folder 9: Speeches; Box 1; NAB Records; WHS.

168. Edwin C. Johnson, "Will Radio-TV Regulate Itself, or Will Gov't Take over Job?" *Variety*, January 2, 1952, 109, ProQuest Entertainment Industry Magazine Archive: *Variety* (1905–2000).

169. "More about the Benton Bill," *NARTB Member Service: Confidential Radio Newsletter*, November 12, 1951, 4.

170. "Benton Bills—A Fair Hearing," *NARTB Member Service: Confidential TV Newsletter*, September 15, 1951, 2.

171. Justin Miller to Frank P. Fogarty, March 10, 1951; Folder 3: J. Miller, March–June 1952; Box 94; NAB Records; WHS.

172. Justin Miller to Harold E. Fellows, June 22, 1951; Folder 8: J. Miller, May–August 1951; Box 93; NAB Records; WHS.

173. NARTB, *Scoreboard*, n/d; Folder 1: NAB Publications: General, Promotional, and Membership Materials, 1940–ca. 1977; Box 118; NAB Records; WHS.

174. William Benton to Dwight D. Eisenhower, November 18, 1959; William Benton Papers; Box 512; Folder 8; SCRC.

175. See William Benton to Dwight D. Eisenhower, draft, November 16, 1959; William Benton Papers; Box 512; Folder 8; SCRC.

176. William Benton to William Hard, November 18, 1959; William Benton Papers; Box 512; Folder 8; SCRC.

177. Dwight D. Eisenhower to William Benton, November 21, 1959; William Benton Papers; Box 512; Folder 8; SCRC.

178. William Benton to Thomas J. Dodd, December 1, 1959; William Benton Papers; Box 512; Folder 8; SCRC.

179. Leverett Saltonstall to William Benton, December 2, 1959; William Benton Papers; Box 512; Folder 8; SCRC.

180. William Benton to John Howe, December 2, 1959; William Benton Papers; Box 512; Folder 8; SCRC.

181. William Benton to Warren G. Magnuson, December 8, 1959; William Benton Papers; Box 512; Folder 8; SCRC.

182. Perlman, *Public Interests*, 18.

Conclusion

1. Semonche, *Censoring Sex*, 194.

2. Hendershot, *Saturday Morning Censors*, 108.

3. Hull, "An Economic Perspective Ten Years After the NAB Case."

4. Ibid., 21.

5. Ibid.

6. Ibid., 31.

7. Ibid., 29.

8. "TV Code Goes into Effect March 1," *Television Digest*, March 1, 1952, 9.

9. "Financial & Trade Notes: Biggest Complaints about TV Code," *Television Digest*, March 29, 1952, 12.

10. Thomas Carskadon and Patrick Murphy Malin to Paul Walker, June 2, 1952; Folder 8: Justin Miller, American Civil Liberties Union, 1945–1953; Box 96; NAB Records; WHS.

11. Ibid.

12. Justin Miller to Bob Richards, September 26, 1952; Folder 4: J. Miller, July–October 1952; Box 94; NAB Records; WHS.

13. Justin Miller to Harold Fellows, October 17, 1952; Folder 4: J. Miller, July–October 1952; Box 94; NAB Records; WHS.

14. Glenn, "'Anything-Goes' TV Era on Way Out," *Sponsor*, November 1951, 82.

15. Ibid.

16. Ibid.

17. Ibid., 81.

18. Justin Miller to Norman R. Glenn, November 13, 1951; Folder 1: J. Miller, November–December 1951; Box 94; NAB Records; WHS.

19. Justin Miller to Norman R. Glenn, December 14, 1951; Folder 1: J. Miller, November–December 1951; Box 94; NAB Records; WHS.

20. Cooper, *Violence on Television*, 11.

21. House Committee on Interstate and Foreign Commerce, "Investigation of Radio and Television Programs," report pursuant to H. Res. 278, 82nd Cong. (1952), 1.

22. Ibid.

23. Ibid.

24. Charles R. Denny to Joseph V. Heffernan and Sylvester Weaver, June 16, 1952; Folder 24: Weaver, 1952, N; Box 121; NBC Records; WHS.

25. House Committee on Interstate and Foreign Commerce, "Investigation of Radio and Television Programs" (1952), 3.

26. Ibid.

27. House Committee on Interstate and Foreign Commerce, subcommittee hearing on H. Res. 278, 82nd Cong. (1952), 393.

28. House Committee on Interstate and Foreign Commerce, "Investigation of Radio and Television Programs" (1952), 12–13.

29. NARTB Television Code Review Board (TCRB), *First Report to the People of the United States* (Washington, DC: NARTB, 1953), 8; Folder 5: NAB Publications, Code Authority, General Material Re: To Radio and Television Code, 1939–ca. 1957 (hereafter Folder 5); Box 122; NAB Records; WHS. Harold Fellows, Ralph Hardy, Thad Brown, and John Fetzer all testified at the hearings.

30. House Committee on Interstate and Foreign Commerce, "Investigation of Radio and Television Programs" (1952), 4.

31. Ibid., 5.

32. Ibid., 4.

33. Ibid., 7.

34. House Committee on Interstate and Foreign Commerce, subcommittee hearing on H. Res. 278, 82nd Cong. 257 (1952).

35. House Committee on Interstate and Foreign Commerce, "Investigation of Radio and Television Programs" (1952), 8.

36. Ibid., 4.

37. Ibid.

38. Ibid., 13.

39. TCRB, *First Report to the People of the United States*, 8; Folder 5; Box 122; NAB Records; WHS.

40. House Committee on Interstate and Foreign Commerce, "Investigation of Radio and Television Programs," (1952), 10.

41. Ibid., 11.

42. Ibid.

43. John E. Fetzer, "Television: To Be or Not to Be!" speech delivered April 30, 1953; Folder 6: Television Code Review Board, 1952–1953; Box 112; NAB Records; WHS.

44. See "Biographical Note," Dorothy Stimson Bullitt Papers, 1933–1993, University of Washington Libraries, Special Collections (website), Archives West, last modified July 17, 2017, archiveswest.orbiscascade.org/ark:/80444/xv26691#bioghistlD.

45. J. Frank Beatty, "Code Review Board," *Broadcasting*, February 18, 1952, 25.

46. "*Television* Magazine Full Market Coverage Circulation," *Television*, November 1953, 51.

47. TCRB, *First Report to the People of the United States*, 15–16; Folder 5; Box 122; NAB Records; WHS.

48. Ibid., 19.

49. Ibid.

50. Ibid., 20.

51. Ibid., 4.

52. Morgan, *The Television Code of the National Association of Broadcasters*, 218.

53. Edward H. Bronson to J. Leonard Reinsch, October 9, 1953; Folder 5: J. Miller, Television Code, 1950–1951; Box 112; NAB Records; WHS.

54. John E. Fetzer, "Television: To Be or Not to Be!" speech delivered April 30, 1953; Folder 6: Television Code Review Board, 1952–1953; Box 112; NAB Records; WHS.

55. Baldwin, Cave, and Lodge, *Understanding Regulation*, 140–141.

56. Morgan, *The Television Code of the National Association of Broadcasters*, 218.

57. John Fetzer to Justin Miller, June 19, 1952; Folder 5: J. Miller, Television Code, 1950–1951; Box 112; NAB Records; WHS.

58. Morgan, *The Television Code of the National Association of Broadcasters*, 218–219.

59. Ibid., 219.

60. Justin Miller to John Fetzer, June 27, 1952; Folder 6: Television Code Review Board, 1952–1953; Box 112; NAB Records; WHS.

61. Morgan, *The Television Code of the National Association of Broadcasters*, 219.

62. Dancer-Fitzgerald-Sample, Inc., to Harold E. Fellows, December 7, 1951; Folder 16: Weaver, 1954, D; Box 123; NBC Records; WHS.

63. House Committee on Interstate and Foreign Commerce, "Investigation of Radio and Television Programs" (1952), 4.

64. Mort Lewis and Morton Wishengrad, "Comments on Provisions of NARTB Code," April 13, 1953; Folder 8: NAB, Justin Miller, American Civil Liberties Union, 1945–1953; Box 96; NAB Records; WHS.

65. Stasheff and Bretz, *The Television Program: Its Writing, Direction and Production*, 75.

66. Morgan, *The Television Code of the National Association of Broadcasters*, 220.

67. Ibid., 221.

68. Available in the Prelinger Archives on the Internet Archive website at https://archive.org/details/WelcomeG1957.

69. The NARTB changed its name back to NAB in 1958, so this film, which is dated ca. 1957, must have been produced concurrently with the name change.

70. Morgan, *The Television Code of the National Association of Broadcasters*, 221.

71. Ibid.

72. Ibid.

73. John E. Fetzer, "The Television Code after Two Years," speech delivered May 26, 1954; HE8689.4.A1 no. 689; Special Collections in Mass Media and Culture Pamphlets; National Association of Broadcasters Collection; LAB. For television stations in operation in May 1954, see "Telestatus," *Broadcasting*, May 31, 1954, 123. The total includes educational stations.

74. Harold E. Fellows, "The Television Code," remarks made on October 15 and October 25, 1956; HE 8689.4.A1 no. 686; National Association of Broadcasters Collection; LAB.

Appendix A

1. NARTB, *The Television Code*, 2–3.

Appendix B

1. Ibid., 3.

Bibliography

Aitken, Hugh G. J. "Allocating the Spectrum: The Origins of Radio Regulation." *Technology and Culture* 35, no. 4 (1994): 686–716.

American Civil Liberties Union (ACLU). *Radio Programs in the Public Interest: Answers to the Radio Industry's Objections to the FCC's New Standards of Judging the Public Service of Radio*. New York: ACLU, 1946. See also Folder 8: NAB Justin Miller, American Civil Liberties Union, 1945–1953; Box 96; NAB Records; WHS.

Anderson, Christopher. *Hollywood TV: The Studio System in the 1950s*. Austin: University of Texas Press, 1994.

Arbuckle, Mark. "Herbert Hoover's National Radio Conferences and the Origins of Public Interest Content Regulation of United States Broadcasting: 1922–1925." PhD diss., Southern Illinois University at Carbondale, 2001.

Aufderheide, Patricia. *Communications Policy and the Public Interest: The Telecommunications Act of 1996*. New York: Guilford Press, 1999.

Baldwin, Robert, Martin Cave, and Martin Lodge. *Understanding Regulation: Theory, Strategy, and Practice*. 2nd ed. Oxford: Oxford University Press, 2012.

Barnouw, Erik. *A Tower of Babel: A History of Broadcasting in the United States to 1933*, vol. 1. New York: Oxford University Press, 1966.

———. *The Golden Web: A History of Broadcasting in the United States, 1933–1953*. New York: Oxford University Press, 1968.

Benjamin, Louise. "Working It Out Together: Radio Policy from Hoover to the Radio Act of 1927." *Journal of Broadcasting and Electronic Media* 42, no. 2 (1998): 221–236.

Besen, Stanley M., Thomas G. Krattenmaker, A. Richard Metzger Jr., and John R. Woodbury. *Misregulating Television: Network Dominance and the FCC*. Chicago: University of Chicago Press, 1984.

Black, Gregory D. *The Catholic Crusade against the Movies, 1940–1975*. Cambridge: Cambridge University Press, 1997.

Boddy, William. *Fifties Television: The Industry and Its Critics*. Urbana: University of Illinois Press, 1993.

Boylan, William A. "Legal and Illegal Limitations on Television Programming." *Journal of the Federal Communications Bar Association* 11, no. 3 (1950): 137–149.

Brenner, Daniel L. "The Limits of Broadcast Self-Regulation under the First Amendment." *Stanford Law Review* 27 (1974–1975): 1527–1562.

Brinson, Susan. *The Red Scare, Politics, and the Federal Communications Commission, 1941–1960*. Westport, CT: Praeger, 2004.

Cassidy, Marsha F. *What Women Watched: Daytime Television in the 1950s*. Austin: University of Texas Press, 2005.

Clarke, Michael. *Regulation: The Social Control of Business between Law and Politics*. New York: St. Martin's Press, 2000.

Cole, Barry, and Mal Oettinger. *Reluctant Regulators: The FCC and the Broadcast Audience*. Reading, MA: Addison-Wesley Publishing Co., 1978.

Cooper, Cynthia A. *Violence on Television: Congressional Inquiry, Public Criticism, and Industry Response: A Policy Analysis*. Lanham, MD: University Press of America, 1996.

Cox, Kenneth A. "The FCC's Role in Television Programming Regulation." *Villanova Law Review* 14 (Summer 1969): 590–601.

Diem, Sarah, Michelle D. Young, Anjalé D. Welton, Katherine Cumings Mansfield, and Pei-Ling Lee. "The Intellectual Landscape of Critical Policy Analysis." *International Journal of Qualitative Studies in Education* 27, no. 9 (2014): 1068–1090.

Doherty, Thomas. *Cold War, Cool Medium: Television, McCarthyism, and American Culture*. New York: Columbia University Press, 2003.

Douglas, Susan. *Inventing American Broadcasting: 1899–1922*. Baltimore: Johns Hopkins University Press, 1987.

Federal Communications Commission (FCC). *Public Service Responsibility of Broadcast Licensees*. Washington, DC: FCC, 1946.

———. *Sixteenth Annual Report*. Washington, DC: US Government Printing Office, 1951.

———. *Seventeenth Annual Report*. Washington, DC: US Government Printing Office, 1952.

Feuer, Jane. "The Concept of Live Television: Ontology as Ideology." In *Regarding Television: Critical Approaches—An Anthology*, edited by E. Ann Kaplan, 12–22. Frederick, MD: University Publications of America, 1983.

———. "Genre Study and Television." In *Channels of Discourse, Reassembled: Television and Contemporary Criticism*, edited by Robert C. Allen, 101–134. 2nd ed. Chapel Hill: University of North Carolina Press, 1992.

Fischer, Frank. *Reframing Public Policy: Discursive Politics and Deliberative Practice*. New York: Oxford University Press, 2003.

———. "What Is Critical? Connecting the Policy Analysis to Political Critique." *Critical Policy Studies* 10, no. 1 (2016): 95–98.

Flannery, Gerald V. *Commissioners of the FCC, 1927–1994*. Lanham, MD: University Press of America, 1995.

Fleming, J. Carlton. "Television Programming: Its Legal Limitations." *Duke Bar Journal* 1, no. 1 (March 1951): 5–25.

Forman, Murray. "Television before Television Genre: The Case of Popular Music." *Journal of Popular Film and Television* 31, no. 1 (2003): 5–16.

Freeman, Jody. "The Private Role in Public Governance." *New York University Law Review* 75, no. 3 (2000): 543–675.

Goldberg, Henry, and Michael Couzens. "'Peculiar Characteristics': An Analysis of the First Amendment Implications of Broadcast Regulation." *Federal Communications Law Journal* 31, no. 1 (Winter 1978): 1–50.

Graham, James M., and Victor H. Kramer. *Appointments to the Regulatory Agencies: The Federal Communications Commission and the Federal Trade Commission, 1949–1974.* Washington, DC: US Government Printing Office, 1976.

Harper, Fowler V., and Edwin D. Etherington. "What the Supreme Court Did Not Do during the 1950 Term." *University of Pennsylvania Law Review* 100, no. 3 (December 1951): 354–409.

Hawkesworth, Mary. "Policy Studies within a Feminist Frame." *Policy Sciences* 27, nos. 2/3 (1994): 97–118.

Hendershot, Heather. *Saturday Morning Censors: Television Regulation before the V-Chip.* Durham, NC: Duke University Press, 1998.

Hills, Matt. *The Pleasures of Horror.* London: Continuum, 2005.

Hilmes, Michele. *Hollywood and Broadcasting.* Urbana: University of Illinois Press, 1990.

Holt, Jennifer. *Empires of Entertainment: Media Industries and the Politics of Deregulation, 1980–1996.* New Brunswick, NJ: Rutgers University Press, 2011.

———. "NYPD Blue: Content Regulation." In *How to Watch Television*, edited by Ethan Thompson and Jason Mittell, 271–280. New York: New York University Press, 2013.

Horwitz, Robert Britt. *The Irony of Regulatory Reform: The Deregulation of American Telecommunications.* New York: Oxford University Press, 1989.

Howarth, David, Jason Glynos, and Steven Griggs. "Discourse, Explanation, and Critique." *Critical Policy Studies* 10, no. 1 (2016): 99–104.

Hull, Brooks B. "An Economic Perspective Ten Years after the NAB Case." *Journal of Media Economics* 3, no. 1 (March 1990): 19–35.

Jaramillo, Deborah L. "Astrological TV: The Creation and Destruction of a Genre." *Communication, Culture, and Critique* 8, no. 2 (2015): 309–326.

———. "Keep Big Government out of Your Television Set: The Rhetoric of Self-Regulation before the Television Code." In *Production Studies, The Sequel!: Cultural Studies of Media Industries*, vol. 2, edited by Miranda Banks, Bridget Conor, and Vicky Mayer, 251–258. New York: Routledge, 2016.

———. "The Rise and Fall of the Television Broadcasters Association, 1943–1951." *Journal of E-Media Studies* 5, no. 1 (2016): 1–33. doi:10.1349/PS1.1938–6060.A.459.

Jassem, Harvey C. "An Examination of Self-Regulation of Broadcasting." *Communications and the Law* 5, no.2 (Spring 1983): 51–64.

Johnson, Nicholas. "A New Fidelity to the Regulatory Ideal." *Georgetown Law Journal* 59 (1970–1971): 869–908.

Killmeier, Matthew A. "More than Monsters: Dark Fantasy, the Mystery-Thriller, and Horror's Heterogeneous History." *Journal of Radio and Audio Media* 20, no. 1 (2013): 165–180.

Kompare, Derek. *Rerun Nation: How Repeats Invented American Television.* New York: Routledge, 2005.

Krasnow, Erwin G., Lawrence D. Longley, and Herbert A. Terry. *The Politics of Broadcast Regulation.* 3rd ed. New York: St. Martin's Press, 1982.

Krattenmaker, Thomas G., and A. Richard Metzger Jr. "FCC Regulatory Authority over Commercial Television Networks: The Role of Ancillary Jurisdiction." *Northwestern University Law Review* 77, no. 4 (November 1982): 403–491.

Krattenmaker, Thomas G., and Lucas A. Powell Jr. *Regulating Broadcast Programming.* Cambridge, MA: MIT Press, 1994.

Landis, James M. *Report on Regulatory Agencies to the President-Elect*. Washington, DC: US Government Printing Office, 1960; Kindle edition, New Orleans, LA: Quid Pro Books, 2014.

Leff, Leonard J. *The Dame in the Kimono: Hollywood, Censorship, and the Production Code*. Lexington: University Press of Kentucky, 2001.

Levin, Harvey J. "The Limits of Self-Regulation." *Columbia Law Review* 67, no. 4 (April 1967): 603–644.

Lichty, Lawrence W. "Members of the Federal Radio Commission and Federal Communications Commission, 1927–1961." *Journal of Broadcasting* 6, no. 1 (Winter 1961–1962): 23–34.

Loevinger, Lee. "The Issues in Program Regulation." *Federal Communications Bar Journal* 20, no. 1 (1966): 3–15.

Lowi, Theodore J. *The End of Liberalism: The Second Republic of the United States*. 2nd ed. New York: W. W. Norton & Co., 1979.

Luke, Timothy W. "What Is Critical?" *Critical Policy Studies* 10, no. 1 (2016): 113–116.

MacCarthy, Mark M. "Broadcast Self-Regulation: The NAB Codes, Family Viewing Hour, and Television Violence." *Cardozo Arts and Entertainment Law Journal* 13 (1995): 667–696.

Mackey, David R. "The National Association of Broadcasters: Its First Twenty Years." PhD diss., Northwestern University, 1956.

Mansfield, Katherine Cumings, Anjalé D. Welton, and Margaret Grogan. "'Truth or Consequences': A Feminist Critical Policy Analysis of the STEM Crisis." *International Journal of Qualitative Studies in Education* 27, no. 9 (2014): 1155–1182.

Martin, Catherine. "You Don't Have to Be a Bad Girl to Love Crime." PhD diss. in progress, Boston University.

McCarthy, Anna. *The Citizen Machine: Governing by Television in 1950s America*. New York: New Press, 2010.

McChesney, Robert W. *Telecommunications, Mass Media, and Democracy: The Battle for the Control of US Broadcasting, 1928–1935*. New York: Oxford University Press, 1993.

McCusker, Kristine M. "'Dear Radio Friend': Listener Mail and the *National Barn Dance*, 1931–1941." *American Studies* 39, no. 2 (Summer 1998): 173–195.

Meehan, Eileen R. "Watching Television: A Political Economic Approach." In *A Companion to Television*, edited by Janet Wasko, 238–255. Malden, MA: Wiley-Blackwell, 2010.

———. "Critical Theorizing on Broadcast History." In *Routledge Reader on Electronic Media History*, edited by Donald G. Godfrey and Susan L. Brinson, 30–44. New York: Routledge, 2015.

Meyer, Michaela D. E. "New Directions in Critical Television Studies: Exploring Text, Audience, and Production in Communication Scholarship." *Communication Studies* 63, no. 3 (July/August 2012): 263–268.

Meyers, Cynthia B. "From Radio Adman to Radio Reformer: Senator William Benton's Career in Broadcasting, 1930–1960." *Journal of Radio and Audio Media* 16, no. 1 (May 2009): 17–29.

Miller, Justin. *The Blue Book: An Analysis by Justin Miller, President, National Association of Broadcasters*. Washington, DC: NAB, 1947. See also Folder 6: Speeches, Justin Miller, 1947; Box 1; NAB Records; WHS.

Morgan, Robert Shepherd. *The Television Code of the National Association of Broadcasters: The First Ten Years*. PhD diss., State University of Iowa, 1964. Ann Arbor, MI: University Microfilms International, 1964.

Mosco, Vincent. *Broadcasting in the United States: Innovative Challenge and Organizational Control*. Norwood, NJ: Ablex Publishing Corp., 1979.

Murray, Matthew. "Television Wipes Its Feet: The Commercial and Ethical Considerations behind the Adoption of the Television Code." *Journal of Popular Film and Television* 21, no. 3 (Fall 1993): 128–138.

——. "'The Tendency to Deprave and Corrupt Morals': Regulation and Irregular Sexuality in Golden Age Radio Comedy." In *Radio Reader: Essays in the Cultural History of Radio*, edited by Michelle Hilmes and Jason Loviglio, 135–156. New York: Routledge, 2002.

Murray, Matthew John. "Broadcast Content Regulation and Cultural Limits, 1920–1962." PhD diss., University of Wisconsin–Madison, 1997.

Murray, Susan. "Our Man Godfrey: Arthur Godfrey and the Selling of Stardom in Early Television." *Television and New Media* 2, no. 3 (August 2001): 187–204.

National Association of Broadcasters (NAB). *Standards of Practice for American Broadcasters*. Washington, DC: NAB, 1948.

National Association of Radio and Television Broadcasters (NARTB). *The Television Code of the National Association of Radio and Television Broadcasters*. Washington, DC: NARTB, 1952.

National Broadcasting Company (NBC). *Responsibility: A Working Manual of NBC Program Policies*. New York: NBC, 1948. See also Folder 3: NBC Library: Program Policies, 1945–1956; Box 220; NBC Records; WHS.

——. *Television Today: Its Impact on People and Products*. New York: NBC Television, 1951. See also File 37-109: Television (A2–Z2); Box 8; Hedges Papers; LAB.

——. *NBC Radio and Television Broadcast Standards*. New York: NBC, 1951. See also Folder 3: NBC Library: Program Policies, 1948–1956; Box 220; NBC Records; WHS.

National Industrial Conference Board. *Trade Associations and Their Economic Significance and Legal Status*. New York: National Industrial Conference Board, 1925.

Neale, Steve. *Genre and Hollywood*. London: Routledge, 2000.

Netzhammer, Emile C., III. "Self-Regulation in Broadcasting: A Legal Analysis of the National Association of Broadcasters Television Code." Master's thesis, University of Utah, 1984.

Noll, Roger G., Merton J. Peck, and John J. McGowan. *Economic Aspects of Television Regulation*. Washington, DC: Brookings Institution, 1973.

Owens, Andrew J. "Coming Out of the Coffin: Queer Historicity and Occult Sexualities on ABC's *Dark Shadows*." *Television and New Media* 17, no. 4 (2016): 350–365.

Perlman, Allison. *Public Interests: Media Advocacy and Struggles over US Television*. New Brunswick, NJ: Rutgers University Press, 2016.

Pickard, Victor. "The Battle over the FCC Blue Book: Determining the Role of Broadcast Media in a Democratic Society, 1945–8." *Media, Culture, and Society* 33, no. 2 (2011): 171–191.

Pondillo, Robert. *America's First Network TV Censor: The Work of NBC's Stockton Helffrich*. Carbondale: Southern Illinois University Press, 2010.

Porst, Jennifer. "*United States v. Twentieth Century–Fox, et al.* and Hollywood's Feature Films on Early Television." *Film History* 25, no. 4 (2013): 114–142.

Prosser, Tony. "Regulation and Social Solidarity." Journal of Law and Society 33, no. 3 (September 2006): 364–387.

Rabin, Robert L. "Federal Regulation in Historical Perspective." *Stanford Law Review* 38, no. 5 (May 1986): 1189–1326.

Robinson, Glen O. "The FCC and the First Amendment: Observations on 40 Years of Radio and Television Regulation." *Minnesota Law Review* 52, no. 1 (November 1967): 67–163.

———. "The Federal Communications Commission: An Essay on Regulatory Watchdogs." *Virginia Law Review* 64, no. 2 (March 1978): 169–262.

Rosenberg, Herbert H. "Program Content: A Criterion of Public Interest in FCC Licensing." *Western Political Quarterly* 2, no. 3 (September 1949): 375–401.

Scarpa, Carlo. "The Anticompetitive Effects of Minimum Quality Standards: The Role of Self-Regulation." In *The Anticompetitive Impact of Regulation*, edited by Giuliano Amato and Laraine L. Laudati, 29–48. Cheltenham, UK: Edward Elgar, 2001.

Schatz, Thomas. *Hollywood Genres: Formulas, Filmmaking, and the Studio System.* Boston: McGraw-Hill, 1981.

Schmidt, Lisa. "Television: Horror's 'Original' Home." *Horror Studies* 4, no. 2 (2013): 159–171

Semonche, John E. *Censoring Sex: A Historical Journey through American Media.* Lanham, MD: Rowman & Littlefield Publishers, 2007.

Sewell, Philip W. *Television in the Age of Radio: Modernity, Imagination, and the Making of a Medium.* New Brunswick, NJ: Rutgers University Press, 2014.

Silverman, David S. *You Can't Air That: Four Cases of Controversy and Censorship in American Television Programming.* Syracuse, NY: Syracuse University Press, 2007.

Simmons, Charlene. "Dear Radio Broadcaster: Fan Mail as a Form of Perceived Interactivity." *Journal of Broadcasting and Electronic Media* 53, no. 3 (September 2009): 444–459.

Spigel, Lynn. *Make Room for TV: Television and the Family Ideal in Postwar America.* Chicago: University of Chicago Press, 1992.

Stasheff, Edward, and Rudy Bretz. *The Television Program: Its Writing, Direction, and Production.* New York: A. A. Wyn, Inc., 1951.

"State and Local Censorship of Films Used on Television." *Federal Communications Bar Journal* 4 (Winter 1949): 193–200.

Sterling, Christopher H., Cary O'Dell, and Michael C. Keith, eds. *The Concise Encyclopedia of American Radio.* New York: Routledge, 2010.

Streeter, Thomas. *Selling the Air: A Critique of the Policy of Commercial Broadcasting in the United States.* Chicago: University of Chicago Press, 1996.

Television Broadcasters Association (TBA). *Proceedings of the First Annual Conference of the Television Broadcasters Association, Inc.* New York: TBA, 1944.

———. *Proceedings of the Second Television Conference and Exhibition, Television Broadcasters' Association, Inc.* New York: TBA, 1946.

———. *Proceedings of the Television Clinic of Television Broadcasters Association, Inc.* New York: TBA, 1949.

Tillinghast, Charles M. *American Broadcasting Regulation and the First Amendment: Another Look.* Ames: Iowa State University Press, 2000.

Tuchman, Gaye, ed. *The TV Establishment: Programming for Power and Profit.* Englewood Cliffs, NJ: Prentice-Hall, 1974.

Turow, Joseph. "Another View of 'Citizen Feedback' to the Mass Media." *Public Opinion Quarterly* 41, no. 4 (Winter 1977–1978): 534–543.

——. "Audience Construction and Cultural Production: Marketing Surveillance in the Digital Age." *Annals of the American Academy of Political and Social Science* 597 (January 2005): 103–121.

US Bureau of the Census. *1950 Census of Population.* Washington, DC: US Department of Commerce, 1951.

Waller, Gregory. *American Horrors: Essays on the Modern American Horror Film.* Urbana: University of Illinois Press, 1987.

Wang, Jennifer Hyland. "Convenient Fictions: The Construction of the Daytime Broadcast Audience, 1927–1960." PhD diss., University of Wisconsin–Madison, 2006.

Wasko, Janet, and Eileen R. Meehan. "Critical Crossroads or Parallel Routes? Political Economy and New Approaches to Studying Media Industries and Cultural Products." *Cinema Journal* 52, no. 3 (Spring 2013): 150–157.

Wilbur, Susan K. "The History of Television in Los Angeles, 1931–1952: Part I: The Infant Years." *Southern California Quarterly* 60, no. 1 (1978): 59–76.

Zarkin, Kimberly, and Michael J. Zarkin. *The Federal Communications Commission: Front Line in the Culture and Regulation Wars.* Westport, CT: Greenwood Press, 2006.

Manuscript Collections

Federal Communications Commission (FCC), Record Group 173. Records. National Archives at College Park, College Park, MD.

National Association of Broadcasters (NAB) Records, 1938–1982. Papers. Wisconsin Historical Society, Madison.

National Broadcasting Company (NBC) History Files, 1922–1986. Papers. Library of Congress, Washington, DC.

National Broadcasting Company (NBC) Records, 1921–2000. Papers. Wisconsin Historical Society, Madison.

William Benton Papers, 1839–1973. Papers. Special Collections Research Center (SCRC), University of Chicago Library, Chicago.

William S. Hedges Papers. Papers. Library of American Broadcasting (LAB), University of Maryland Library, College Park.

Index

Italic page numbers indicate material in figures.